Multiobjective Optimization: Behavioral and Computational Considerations

Multiobjective Optimization: Behavioral and Computational Considerations

Jeffrey L. Ringuest
Operations Research and Strategic Management
Wallace Carroll School of Management
Boston College
Chestnut Hill, MA 02167-3808
USA

SPRINGER SCIENCE+BUSINESS MEDIA, LLC

Library of Congress Cataloging-in-Publication Data

Ringuest, Jeffrey L.
    Multiobjective optimization : behavioral and computational
considerations / Jeffrey L. Ringuest.
        p.   cm.
    Includes bibliographical references and index.
    ISBN 978-0-7923-9236-1   ISBN 978-1-4615-3612-3 (eBook)
    DOI 10.1007/978-1-4615-3612-3
    1. Programming (Mathematics)   2. Multiple criteria decision
making.   3. Mathematical optimization.   I. Title.
QA402.5.R56   1992
 658.4'033--dc20   92-6353
                                                                     CIP

To:

J. R.

W. L.

J. R.

TABLE OF CONTENTS

PREFACE

Throughout the development of mathematical programming researchers have paid great attention to problems that are described by a single objective that can only be achieved subject to satisfying a set of restrictions or constraints. Recently, it has been recognized that the use of a single objective limits the applicability of mathematical programming models. In reality, many multiobjective situations exist and frequently these multiple objectives are in direct conflict.

Research on multiobjective problems can be broken down into two broad categories: multiobjective optimization and multicriterion decision theory. Multiobjective optimization models are based on techniques such as linear programming.

In general, the multiobjective optimization problem can be defined as finding a feasible alternative that yields the most preferred set of values for the objective functions. This problem differs from a single objective because subjective methods are required to determine which alternative is most preferred.

A body of literature parallel to that in multiobjective optimization has been developing in the area of multicriterion decision theory. These models are based on classical decision analysis, particularly utility theory. One focus of this research has been the development and testing of procedures for estimating multiattribute utility functions that are consistent with rational decision maker behavior. A utility function provides a model of a decision maker's choice among alternatives. This literature is directly applicable to multiobjective optimization and provides much needed insight into the subjective character of that problem.

Although the commonalities between multiobjective optimization and multicriterion decision theory are obvious, the work is far from unified. Rosenthal ["Principles of Multiobjective Optimization," *Decision Sciences*, 1985, 16, 133-152] suggests that combining multiobjective optimization with multicriterion decision making, "... offers a promising, but much overlooked, approach."

Chapters 1 through 5 of this book, explore this approach, by unifying many of the concepts of multiobjective optimization with multicriterion decision theory. Chapter 1 provides the background material in multiobjective optimization and multiattribute utility theory necessary to support subsequent chapters. Chapters 2 and 3 discuss the goal programming model and it's implications for the behavior of the decision maker from the perspective of multiattribute utility theory. A more general goal programming model is then presented. Compromise programming and its relationship to goal programming and multiattribute value theory are discussed in Chapter 4. Chapter 5 describes methods for generating the set of nondominated solutions to a multiobjective optimization problem and discusses methods for choosing a solution from among the nondominated set. Interactive methods are discussed in Chapter 6 with the emphasis on the types of tradeoff information which can be elicited from the decision maker.

While the primary goal of this book is to unify the concepts of multicriterion decision theory with those of multiobjective optimization, a second goal is to examine computational issues related to multiobjective optimization. Algorithms for solving multiobjective optimization problems typically require much more storage and CPU time than those for single objective problems. In addition, many algorithms require an interaction between the decision maker and the computer. This suggests that computational efficiency is extremely important in multiobjective optimization problems. Chapter 7 of this book describes computational advantages which can be gained in multiobjective optimization problems with a particular structure (network flow problems), while Chapter 8 discusses computational issues in problems of more general structure.

The last two chapters of this book, Chapter 9 and Chapter 10, describe a particular application of multiobjective optimization and framing effects in multiobjective optimization. Chapter 9 shows how multiobjective optimization can be used as an alternative to maximizing net present value which is much less restrictive in terms of the decision maker's preferences over time. Chapter 10 shows how the form of data representation (i.e. the way data is displayed and especially the way data is phrased or framed) may influence the decision maker and gives the implications of data representation on multiobjective optimization methods.

The principle uniqueness of this book is in the strong links drawn between multiattribute decision theory and multiobjective optimization. This book is not a survey of the field of multiobjective optimization. Instead it fills the gap between multiobjective optimization and multiattribute decision theory and draws the attention of researchers in multiobjective optimization to the decision theory literature. In addition, this book presents some of the first discussion of the important computational issues in multiobjective optimization as well as some of the first results of computational testing.

My research in multiobjective optimization has been influenced by many people. Principle among them are the people with whom I have coauthored papers. My thanks go to Joanna Baker, Jonathan Bard, Cengiz Haksever, Mark McKnew, Mick Peters, Dan Rinks and especially Sam Graves and Tom Gulledge.

Multiobjective Optimization: Behavioral and Computational Considerations

INTRODUCTION

1.1 MULTIPLE-OBJECTIVE OPTIMIZATION

Since the end of World War II operations research has developed as a field of study dealing with applications of the scientific method to business operations and management decision making. Throughout its development operations researchers have paid great attention to problems that are described by a single objective that can only be optimized subject to a set of restrictions or constraints. The class of mathematical models that are used to represent such problems are termed mathematical programs. Mathematical programming algorithms are perhaps the most important group of available quantitative techniques in operations research.

The general single-objective, $Z(\mathbf{x})$, mathematical program with n decision variables, x_j, and m constraints, $g_i(\mathbf{x})$, may be defined mathematically as:

Optimize: $Z(x_1, x_2, ..., x_n)$

Subject to: $g_i(x_1, x_2, ..., x_n) \leq 0;$

$x_j \geq 0;$

$i = 1, 2, ..., m;$

$j = 1, 2, ..., n$

or in vector notation

Optimize: $Z(\mathbf{x})$

Subject to: $\mathbf{g}(\mathbf{x}) \leq \mathbf{0}$

$\mathbf{x} \geq \mathbf{0}.$

More recently, it has been recognized that the use of a single objective limits the applicability of models such as mathematical programs. In reality, many multiple-objective situations exist. For example, in government, budget dollars are allocated to accomplish multiple-objectives in areas such as health, education and the military. In business, firms wish to achieve desirable levels of market share, return on investment, employment stability, etc.   When multiple-objectives exist they are frequently in direct conflict. For instance, the objective of maximizing return on investment may well be in conflict with the objective of maintaining a stable work force. Rosenthal (1985) provides a very thorough review of multiobjective optimization.   Much of the remainder of this section and the next section parallels Rosenthal's review.

The general k-objective, $Z(x)$, mathematical program with n decision variables, x_j, and m constraints, $g_i(x)$, may be defined mathematically as:

Optimize: $[Z_1(x), Z_2(x), ..., Z_k(x)]$
                  $= Z(x)$
Subject to: $g(x) \leq 0$
                  $x \geq 0.$

In this model the objective, "Optimize $Z(x)$," lacks clear meaning because the set $\{Z(x)\}$ for all feasible x lacks a natural ordering whenever $Z(x)$ is vector-valued. Without a natural ordering, and given two feasible alternatives x_1 and x_2, it may not be possible to definitively determine whether $Z(x_1)$ is "greater than," "less than," or "equal to" $Z(x_2)$. Thus, analyzing the multiple-objective mathematical program is not a typical optimization problem.

Regardless of how precisely $Z(x)$ and $g(x)$ are defined, any meaningful and useful definition of multiple-objective optimization requires the incorporation of subjectivity. Specifically, the ordering needed to "Optimize $Z(x)$" is determined by the specific decision maker(s). Two decision makers could very easily produce different rank orderings for an identical set of outcomes $\{Z(x)\}$. The optimal $Z(x)$ is the highest ranked or most preferred outcome as determined by the particular decision maker(s). Thus, it is more meaningful to define the multiple-objective optimization problem as finding a feasible alternative, x, that yields the most preferred vector of objective function values, $Z(x)$. This definition is still somewhat fuzzy since no single model can represent the preference ordering of all decision makers. Further, the inherent ambiguity in the definition of the optimization problem makes it difficult to specify the properties desired in an optimal solution.

1.2 DOMINANCE AND EFFICIENCY

One property that is commonly considered as necessary for any candidate solution to the multiple-objective optimization problem is that the solution is not dominated. This property requires the assumption of increasing monotonicity. That is, for every objective function, Z_h (which may be redefined as $-Z_h$), it is assumed that more of Z_h is always preferred to less of Z_h when all other objectives are held at constant levels. This is very often a reasonable assumption, however, exceptions do exist (e.g. objectives which measure; ph level, sugar content, population level, light or sound intensity, etc.). These exceptions will be addressed, at least indirectly, in later sections.

To define dominance then, a feasible point x_1 is dominated by a second feasible point x_2 if and only if

$$Z_h(x_2) \geq Z_h(x_1); \quad h = 1, 2, ..., k$$

and

$$Z_h(x_2) > Z_h(x_1); \quad \text{for at least one h.}$$

It is now possible to define efficiency. A feasible solution, x_j, is efficient if and only if there is no other feasible solution which dominates it. Clearly, if all Z_h are monotone, a dominated solution will never yield the most preferred vector of objective function values.

Geoffrion (1968) makes a distinction between efficient and properly efficient solutions. He has shown that no candidate solution for the multiple-objective optimization problem may be dominated, but that all efficient solutions may not be reasonable candidates. He qualifies those efficient solutions which should be considered as "properly efficient." If x_1 and x_2 are feasible solutions, then x_1 is properly efficient if the ratio

$$\frac{Z_a(x_2) - Z_a(x_1)}{Z_b(x_1) - Z_b(x_2)}$$

is bounded from above. This ratio is the improvement in the ath objective divided by the decrement in the bth objective due to a change of solution from x_1 to x_2. If this ratio is not bounded from above, an extremely large improvement in Z_a would result from an extremely small decrement in Z_b. A rational decision maker would never refuse such an exchange. Thus, one would always choose a solution which is properly efficient.

In practice, the distinction between proper efficiency and efficiency is not significant. For the special case when the objectives and constraints are all linear functions, Benson and Morin (1977) have shown that all efficient points are properly efficient. Soland (1979) discusses other cases in which improperly efficient points do not exist or are not likely to be considered by the decision maker. Geoffrion (1968) suggests that the analyst unsure about the proper efficiency of a point should test numerous values of the ratio about the selected point. Thus, throughout the remaining discussion it will be assumed that the analyst has followed one of these paths and the terms efficient and efficiency will be used to refer to properly efficient points.

This discussion of efficient solutions has not shed any light on the subjectivity inherent in the problem. Other techniques are needed to model this subjectivity. Behavioral decision theory is the area of operations research that has investigated these problems, although in a somewhat different context. An exploration of this area follows below.

1.3 MULTIATTRIBUTE VALUE AND UTILITY THEORY

Managerial decisions often involve tradeoffs among conflicting objectives. For example, choosing the location for a waste treatment plant involves tradeoffs among economic environmental and political objectives. In these situations, the decisions require that managers assess a set of multiattribute alternatives. Mathematically, the decision problem can be posed as follows. Each alternative is represented by a k-dimensional vector $\mathbf{A} = (A_1, A_2, ..., A_k)$ where each A_h characterizes a particular alternative along k relevant attributes. The decision maker seeks to identify the most preferred alternative from among the set {A} of vectors representing all possible alternatives. In doing so it is presumed that the decision maker is guided by an underlying construct that represents his/her preference structure. If the manager's preference structure satisfies certain assumptions, it is possible to represent his/her preferences by a real-valued utility function $U(\mathbf{A})$. The utility function is then used to determine the preference order for a set {A} since $U(\mathbf{A})$ can be defined such that $U(A_1) > U(A_2)$ if and only if A_1 is preferred to A_2.

An important practical issue in applying an analytical decision procedure is in determining the form of the multiattribute utility function. One approach to the problem is to explicitly test for various independence conditions. The results of these tests identify the form of the underlying utility function. A second approach is to fit a simplified function under the assumption that it is the underlying utility function or at least that it will reasonably approximate the underlying true utility function.

1.4   FUNCTIONAL FORMS AND INDEPENDENCE CONDITIONS

Keeney and Raiffa (1976) have developed the most commonly applied procedure for assessing multiattribute utility functions. Using their procedure the multiattribute alternative is first decomposed into its constituent attributes. A conditional utility function is then determined for each attribute. These functions are conditional in the sense that all other attributes are assumed to be fixed at known levels. Finally, the conditional utility functions are combined into a composite utility function. In order to make the assessments that are required to combine the attribute utilities, the form of the composite utility function must be known.

Several decomposition forms of the multiattribute utility function, $U(A)$, have been developed. These include the 1. additive, 2. multiplicative and 3. multilinear forms:

1. Additive form

$$U(A) = \sum_{h=1}^{k} \alpha_h U_h(A_h)$$

2. Multiplicative form

$$1 + \alpha U(A) = \prod_{h=1}^{k} [1 + \alpha \alpha_h U_h(A)]$$

3. Multilinear form

$$U(A) = \sum_{h=1}^{k} \alpha_h U_h(A_h)$$

$$+ \sum_{h=1}^{k} \sum_{h'>h} \alpha_{hh'} U_h(A_h) U_{h'}(A_{h'})$$

$$+ \sum_{h=1}^{k} \sum_{h'>h} \sum_{h''>h'} \alpha_{hh'h''} U_h(A_h)$$

$$U_{h'}(A_{h'}) U_{h''}(A_{h''})$$

$$+ \dots + {}_{123\dots k} U_1(A_1)$$

$$U_2(A_2) \dots U_k(A_k)$$

In each of these cases:

$0 < U_h(A_h) < 1$ and $U_h(A_h)$ is a conditional single attribute utility function such that $U_h(A_h^0) = 0$ and $U_h(A_h^*) = 1$ for some A_h^0 and A_h^*;

$0 < \alpha_h, \alpha_{hh'}, \dots, \alpha_{hh'\dots k} < 1$ are scaling constants;

$-1 < \alpha \le \infty$ and α is a parameter.

1.4.1 Multilinear Utility Function

Keeney and Raiffa (1976) have determined the assumptions which underlie these decomposition forms. The multilinear model is the most general and as such requires the least restrictive independence assumptions. To specify the assumptions of this model and the others, it is first necessary to define utility independence. An attribute A_h is utility independent of its complement $(A_1, A_2, ...,$ $A_{h-1}, A_{h+1}, ..., A_k)$ if the conditional preference order for levels of A_h does not depend on the levels of attributes $(A_1, A_2, ..., A_{h-1},$ $A_{h+1}, ..., A_k)$. The multilinear model is appropriate if and only if each attribute is utility independent of its complement.

To illustrate the independence assumptions, consider the process of choosing a new house. For simplicity, let each house be described by three attributes: selling price, interior square footage and distance from work. Then, to demonstrate utility independence, assume that the decision maker prefers the upper gamble to the lower gamble in the following pair

```
       ┌─(100K, 2000 ft.2, 20 mi.)
       │ p
       │
  ─────┤
       │ 1-p
       └─(75K, 2000ft.2, 20 mi.)
```
            >-

$$
\begin{array}{l}
\rule[-1.5em]{0.5pt}{2em}\!\!\!-(150K, 2000\ ft.^2, 20\ mi.) \\
\quad p' \\
\\
\quad 1\text{-}p' \\
\rule{0.5pt}{0pt}\!\!\!-(50K, 2000\ ft.^2, 20\ mi.)
\end{array}
$$

where >- is read "is preferred to." The upper gamble has a probability, p, of obtaining a house that costs $100,000, has 2000 square feet of interior space and is 20 miles from work; and a probability, 1-p, of obtaining a house that costs $75,000, has 2000 square feet of interior space and is 20 miles from work. The lower gamble has a probability, p', of obtaining a house that costs $150,000, has 2000 square feet of interior space and is 20 miles from work; and a probability, 1-p', of obtaining a house that costs $50,000, has 2000 square feet of interior space and is 20 miles from work. Selling price would be utility independent of the other two attributes if and only if the following relationship also holds for all values of δ_1 and δ_2.

$$
\begin{array}{l}
\rule{0.5pt}{0pt}\!\!\!-[\$100K, (2000+\delta_1)\ ft.^2, (20+\delta_2)\ mi.] \\
\quad p \\
\\
\quad 1\text{-}p \\
\rule{0.5pt}{0pt}\!\!\!-[\$75K, (2000+\delta_1)\ ft.^2, (20+\delta_2)\ mi.] \\
\qquad\qquad >\text{-} \\
\rule{0.5pt}{0pt}\!\!\!-[\$150K, (2000+\delta_1)\ ft.^2, (20+\delta_2)\ mi.] \\
\quad p \\
\\
\quad 1\text{-}p \\
\rule{0.5pt}{0pt}\!\!\!-[\$50K, (2000+\delta_1)\ ft.^2, (20+\delta_2)\ mi.]
\end{array}
$$

1.4.2 Multiplicative Utility Function

The multiplicative model is a more restrictive special case of the multilinear model. This decomposition form requires mutual utility independence. For attributes A_1, A_2, ..., A_k to be mutually utility independent every subset of $\{A_1, A_2, ..., A_k\}$ must be utility independent of its complement.

To illustrate mutual utility independence, suppose that the upper lottery is preferred to the lower lottery:

```
  ┌─($100K, 2000 ft.², 20 mi.)
  │
  │ p
  │
──┤
  │
  │ 1-p
  └─($75K, 1500ft.², 20 mi.)

            >-

  ┌─($150K, 2200 ft.², 20 mi.)
  │
  │ p'
  │
──┤
  │
  │ 1-p'
  └─($50K, 1400 ft.², 20 mi.)
```

Selling price and square footage will be mutually independent of distance from work if and only if the following preference also holds for all values of δ.

—[$100K, 2000 ft.^{2}, $(20+\delta)$ mi.]

p

1-p
—[$75K, 1500ft.^{2}, $(20+\delta)$ mi.]

>-

—[$150K, 2200 ft.^{2}, $(20+\delta)$ mi.]

p'

1-p'
—[$50K, 1400 ft.^{2}, $(20+\delta)$ mi.]

1.4.3 Additive Utility Function

The most restrictive assumptions are required by the additive decomposition model. This form is a special case of both the multiplicative and multilinear models. Additive utility independence is a prerequisite of the additive utility model. Attributes A_1, A_2, ..., A_k are said to be additive utility independent if preferences for lotteries on A_1, A_2, ..., A_k depend only on their marginal probability distributions and are not dependent on their joint probability distribution.

Additive utility independence can be illustrated by the following relationship:

```
    ┌─($100K, 2000 ft.², 20 mi.)
    │ 0.5
    │
  ──┤
    │ 0.5
    └─($75K, 1500ft.², 20 mi.)
```

~

```
    ┌─($100K, 1500 ft.², 20 mi.)
    │ 0.5
    │
  ──┤
    │ 0.5
    └─($75K, 2000 ft.², 20 mi.)
```

where ~ is read "is indifferent to." These two gambles have the same marginal probability of obtaining a house that costs $100,000 or $75,000, has 2000 square feet or 1500 square feet of living area, and is 20 miles from work. Of course these examples illustrate only one of each of the types of independence relationships. These relationships must hold for all combinations of attributes as described in the earlier definitions.

1.5   VALUE FUNCTIONS AS COMPARED TO UTILITY FUNCTIONS

The theory of utility has been developed for the case of decision making under uncertainty. This is illustrated by the use of lotteries for uncertain outcomes in the examples of the prior section. Multiattribute value theory is analogous to utility theory but has been developed for cases where the decision is among multiattribute outcomes that can be obtained with certainty.

Decomposition methods like those for multiattribute utility functions are commonly used for assessing value functions. With value functions, however, the assumptions are somewhat less restrictive. Because of this less restrictive nature, the most common decomposition forms are the additive and multiplicative.

1.5.1 Multiplicative Value Function

The multiplicative value function requires the assumption of preference independence. An attribute A_h is preference independent of its complement $(A_1, A_2, ..., A_{h-1}, A_{h+1}, ..., A_k)$ if the preference between two alternatives which differ only in attribute A_h is dependent on the levels of A_h obtained with each alternative and is independent of the levels of its complementary attributes (all of these attribute levels are common to each alternative. For the multiplicative model to be appropriate each attribute must be preference independent of its complement.

1.5.2 Additive Value Function

The additive value model is a more restrictive special case of the multiplicative model. This decomposition form requires mutual preference independence. For attribute $A_1, A_2, ..., A_k$ to be mutually preference independent every subset of $\{A_1, A_2, ..., A_k\}$ must be preference independent of its complement.

1.6   OPTIMIZING THE MULTIATTRIBUTE UTILITY OR VALUE FUNCTION

The theory of multiattribute utility and value would seem to provide an answer to the question of what to optimize in a multiobjective optimization problem. One might first, define the k-dimensional vector of attributes A as $A = [A_1 = Z_1(x), A_2 = Z_2(x), ...,$ $A_k = Z_k(x)] = Z(x)$. Then, if the decision maker's preference structure satisfies certain assumptions as defined in the previous sections, it is possible using decomposition techniques to represent these preferences by a real-valued utility function $U[Z(x)]$ or a real-valued value function $V[Z(x)]$. The value function will be appropriate if the objective functions $Z(x)$ are known with certainty (a common assumptions of mathematical programming); otherwise, the utility function will be appropriate. Finally, the multiobjective optimization problem can be written as:

Optimize: $U[Z(x)]$ or $V[Z(x)]$

Subject to: $g(x) \leq 0$

                          $x \geq 0.$

While this approach might seem straightforward there are several reasons why it has not been widely applied. The assessment of a multiattribute utility or value function is not an easy task. It requires a highly skilled analyst and a cooperative and patient decision maker. The conditional single-attribute utility functions are usually nonlinear, often involving exponential terms. The composite form can be additive, multiplicative or multilinear as described earlier. Thus, the resulting function is frequently highly nonlinear. Nonlinear programming problems certainly can be solved, but software for solving nonlinear problems is not as readily available as is software for linear programming. In addition, the nonlinear programming model is less conducive to sensitivity analysis. Perhaps, however, the primary reason why this approach is not widely used is that the fields of multiobjective optimization and multiattribute decision theory have developed nearly independently.

In spite of the lack of cases in which the multiattribute utility or value function is assessed and optimized, this approach is extremely important. The model which is developed by this approach is the model which underlies all multiobjective optimization. As such, it can not be ignored. All solution techniques for the multiobjective optimization problem must be judged by this model. That is, the behavioral assumptions implicit in any solution methodology must be considered. These methodologies should be consistent with rational decision making behavior.

1.7   REFERENCES

Benson, H. P. and T. L. Morin, "The Vector Maximization Problem: Proper Efficiency and Stability," *SIAM Journal on Applied Mathematics*, 32 (1977), 64-72.

Geoffrion, A. M., "Proper Efficiency and the Theory of Vector Maximization," *Journal of Mathematical Analysis and Applications*, 22 (1968), 307-322.

Keeney, R. L. and H. Raiffa, *Decisions with Multiple Objectives: Preferences and Value Tradeoffs*, John Wiley and Sons, New York, 1976.

Rosenthal, R. E., "Principles of Multiobjective Optimization," *Decision Sciences* 16 (1985), 133-152.

Soland, R. M., "Multicriteria Optimization: A General Characterization of Efficient Solutions," *Decision Sciences*, 10 (1979), 26-38.

1.8   OTHER RELEVANT READINGS

Texts:

Chankong, V. and Y. Y. Haimes, *Multiobjective Decision Making: Theory and Methodology*, North-Holland, New York, 1983.

Goicoechea, A., D. R. Hansen and L. Duckstein, *Multiobjective Decision Analysis with Engineering and Business Applications*, John Wiley and Sons, New York, 1982.

Hwang, C. L. and A. S. M. Masud, *Multiple Objective Decision Making - Methods and Applications*, Springer-Verlag, Berlin, 1979.

Steuer, R. E., *Multiple Criteria Optimization*, John Wiley and Sons, New York, 1985.

Yu, P. L., *Multiple-Criteria Decision Making*, Plenum Press, New york, 1985.

von Winterfeldt, D. and W. Edwards, *Decision Analysis and Behavioral Research*, Cambridge University Press, Cambridge, England, 1986.

Zeleny, M., *Multiple Criteria Decision Making*, McGraw-Hill, New York, 1982.

The following are examples of explicit optimization of multiattribute value or utility functions:

Baker, J., M, McKnew, T. R. Gulledge, Jr. and J. L. Ringuest, "An Application of Multiattribute Utility Theory to the Planning of Emergency Medical Services," *Socio-Economic Planning Sciences*, 18 (1984) 273-280.

Golabi, K., C. W. Kirkwood and A. Sicherman, "Selecting a Portfolio of Solar Energy Projects Using Multiattribute Preference Theory," *Management Science*, 27 (1981), 174-189.

Gros, J., "Power Plant Siting: A Paretian Environmental Approach," *Nuclear Engineering and Design*, 34 (1975), 281-292.

Harrison, T. P. and R. E. Rosenthal, "A Multiobjective Optimization Program foe Scheduling Timber Harvests on Nonindustrial Private Forest Lands," *TIMS Special Study on Systems Analysis in Forestry and Forest Industries*, TIMS Studies in Management Science (Vol. 21), North- Holland, New York, 1986.

Keefer, D. L., "Applying Multiobjective Decision Analysis to Resource Allocation Planning Problems." In S. Zionts (ed.), *Multiple Criteria Problem Solving: Proceedings, Buffalo, N.Y. (USA), 1977*, Springer-Verlag, New York, 1978.

Ringuest, J. L. and T. R. Gulledge, Jr., "A Preemptive Value-Function Method Approach for Multiobjective Linear Programming Problems," *Decision Sciences* 14 (1983), 76-86.

Rinks, D. B., J. L. Ringuest and M. H. Peters, "A Multivariate Utility Function Approach to Stochastic Capacity Planning," *Engineering Costs and Production Economics*, 12 (1987), 3-13.

LINEAR GOAL PROGRAMMING

2.1 THE GOAL PROGRAMMING MODEL

Suppose that a manager has identified a problem that can be formulated as a traditional linear programming problem with one added complication -- the decision being modeled must be judged on the basis of more than one criterion. Now let $Z_1(\mathbf{x})$, $Z_2(\mathbf{x})$, ..., $Z_h(\mathbf{x})$, ..., $Z_k(\mathbf{x})$ model the criteria as linear objective functions, $g(\mathbf{x})$ be a set of linear constraints and restrict all x to be nonnegative. For ease of explanation let's further suppose that the manager can reduce the set of criteria down to three, $k=3$, and that the objective functions can be specified with certainty. Now considering each criterion separately the manager expresses the following preferences:

1. for all feasible **x** more Z_1 is always preferred to less Z_1 and over this range each additional unit of Z_1 adds approximately the same value;

2. for all feasible **x** which yield $Z_2 \leq b_2$ more Z_2 is always preferred to less Z_2 and over this range each additional unit of Z_2 adds approximately the same value; for all feasible **x** which yield $Z_2 > b_2$ all levels of Z_2 are equally preferred and over this range each additional unit of Z_2 adds no additional value;

3. for all feasible **x** which yield $Z_3 \leq b_{3L}$ more Z_3 is always preferred to less Z_3 and over this range each additional unit of Z_3 adds approximately the same value; for all feasible **x** which yield Z_3 in the range $b_{3L} < Z_3 < b_{3R}$ all levels of Z_3 are equally preferred

and over this range each additional unit of z_3 adds no additional value; for all feasible x which yield $Z_3 \geq b_3$ less Z_3 is always preferred to more Z_3 and over this range each additional unit of Z_3 causes the same reduction in value.

These statements of preference are illustrated graphically in Figure 2.1.

Finally, the manager indicates a willingness to make tradeoffs among all three objectives. The manager goes on to state that the difference in preference between the vectors (Z_1', Z_2, Z_3) and (Z_1'', Z_2, Z_3), where these two vectors differ only in the level of Z_1, will be the same regardless of the levels of Z_2 and Z_3. The manager also feels comfortable making an analogous statement for every subset of $\{Z_1, Z_2, Z_3\}$. Thus, the assumption of mutual preference independence, as described in Chapter 1, holds implying an additive multiattribute value function.

This information can be translated into the following goal programming problem:

Minimize: $w_1 p_1 + w_2 p_2 + w_3 p_3 + w_4 n_4$

Subject to: $Z_1 - n_1 + p_1 = b_1$

$Z_2 - n_2 + p_2 = b_2$

$Z_3 - n_3 + p_3 = b_{3L}$

$Z_3 - n_4 + p_4 = b_{3R}$

$n_i, p_i \geq 0; \ i = 1, 2, 3, 4.$

where n_i is the negative deviation from b_i, p_i is the positive deviation from b_i, b_1 is an aspiration level for objective Z_1 and the w's are

Figure 2.1

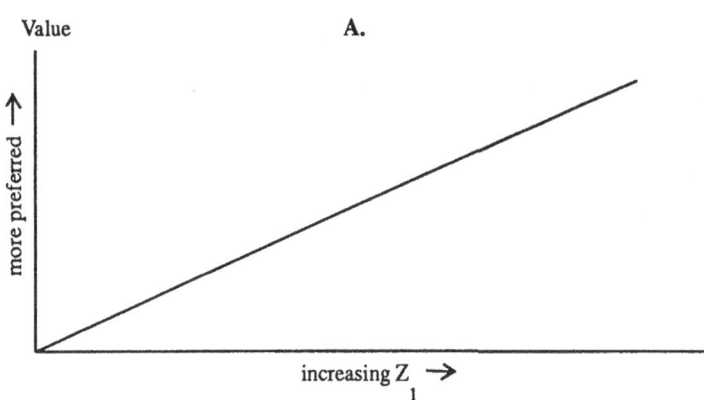

A.

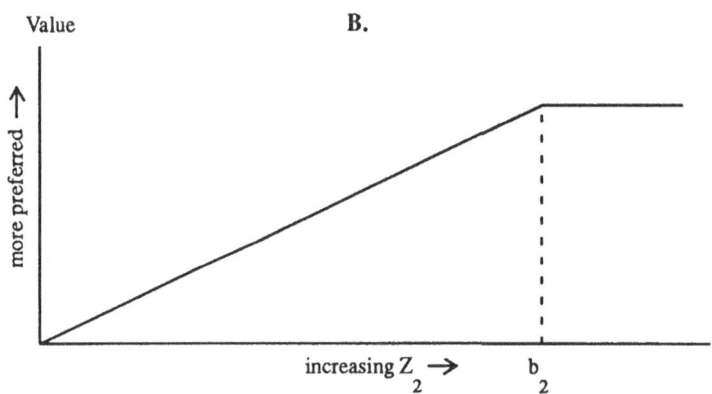

B.

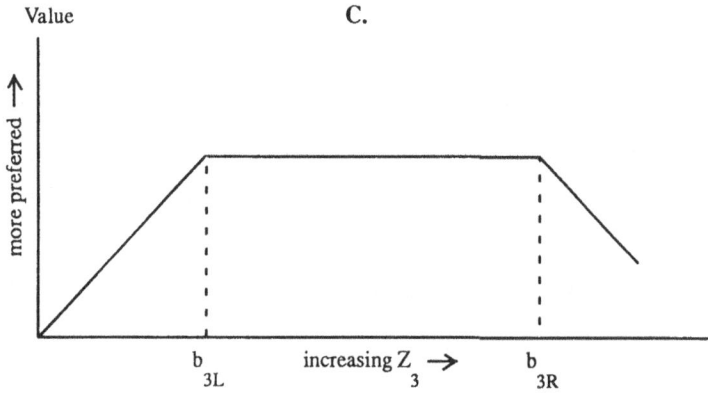

C.

weights used to rate the importance of minimizing each deviational variable.

If the conditions that have been described are satisfied, two additional questions still exist. What values represent the manager's aspiration levels, b's, for each goal and what weights, w's, will model the manager's tradeoffs among objectives?

2.2   ASPIRATION LEVELS

In determining aspiration levels b_1, b_2, b_{3L} and b_{3R}; determining b_1 can be the most problematic. For cases of the form of Figure 2.1A, the manager's goal is in effect to maximize the objective function. The manager's aspiration level should represent this objective. A misspecification of this level can result in a dominated solution. This will occur if the solution obtained meets the specified aspiration level (in this case setting $p_1 = 0$) but higher levels of the objective, Z_1, could be obtained without adversely affecting the other objectives, Z_2 or Z_3. The danger is in setting the aspiration level too low. But what level is high enough?

To be safe, the aspiration level can be set equal to the maximum objective function level. This can be found by solving the single objective linear program for x^* which maximizes Z_1 subject to $g(x) \leq 0$ and $x \geq 0$, assuming of course that this problem is bounded. In this case, $b_1 = Z_1(x^*)$. If the single objective problem is not bounded the aspiration level b_1 should be set at a high enough level so that p_1 is not equal to zero. This may require adjusting the level of b_1 and resolving if the initial guess yields p_1 equal to zero. How this numerical value effects w_1 will be discussed later.

For the second objective above shown in Figure 2.1B, the manager's aim is to achieve some minimum level for Z_2 beyond which there is no benefit. This minimum level specifies the appropriate aspiration level, b_2. The actual level in this case would be imposed by the specific problem application.

The manager's objective in the third case shown in Figure 2.1C is to attain a level for Z_3 that is within some most preferred range. The low end of this range specifies b_{3L} while the upper end specifies b_{3R}. In some instances there may be a single most preferred level for the objective. As a result $b_{hL} = b_{hR}$. Again, the actual levels would be imposed by the specific problem application.

2.3 WEIGHTS

Once the aspiration levels are specified, the goal constraints can be formulated. Then the objective can be formulated as a weighted linear combination of the relevant deviational variables. The weights represent the decision maker's rating of the importance of deviations from the aspiration levels.

Care must be taken in developing these weights. The numerical value of the deviation from each unsatisfied goal is the value of the positive or negative deviational variable in the goal constraint. This numerical value is dependent on the scale associated with the variable. The scale is determined by the units of measure and the specification of the aspiration level. To make the scales equal, deviations of equal geometrical distances must yield equal numerical values (i.e. the scales must be normalized).

Normalization can be accomplished by associating the same Euclidean distance measure with each deviational variable. Goal variables for the hth goal and the hth aspiration level should be weighted by the reciprocal of the Euclidean norm $\|c_h\| = \sqrt{((c_h) \cdot c_h)}$ where c_h represents the vector of coefficients of the decision variables in the hth goal. For example, the goal constraint

$$c_{1h}x_1 + \dots + c_{nh}x_n + n_h - p_h = b_h$$

would become

$$(c_{1h}/\|c_h\|)x_1 + \dots + (c_{nh}/\|c_h\|)x_n$$
$$+ n_h - p_h = b_h/\|c_h\|$$

before the weights are assigned to n_h and p_h in the objective function. The decision maker would determine objective function weights by considering the importance of each goal relative to the importance of the other goals.

2.4   PREEMPTIVE PRIORITIES

An alternative to minimizing the weighted sum of the deviational variables is to assign priorities to the goals and then to optimize them one at a time in order of priority. At each successive optimization, higher priority deviational variables are constrained to their optimum level as determined in previous optimizations.

The purpose of this approach is to determine the best attainable value at the highest priority level. If the best value is attained at a unique solution the process ends. If there are alternative optimal solutions, the alternative which provides the best value at the second priority level is chosen. If ties still remain, the third priority is considered, and so on, until all priority levels are considered or a unique solution is obtained.

One advantage of this approach is that goals measured in different units can be placed at different priority levels and it will not be necessary to determine relative weights. A limitation is that the optimization process will only pass to a lower priority level if there are alternative optimal solutions at the higher priority level. Even more troublesome is that tradeoffs between priority levels are disallowed. Thus, a large improvement in a lower priority objective can not be accepted if a higher priority objective is degraded no matter how minute the degradation is in the higher level objective.

Sherali (1982) and Sherali and Soyster (1983) have demonstrated an interesting relationship between the preemptive priority and weighted summed deviation approaches.   They show that if a preemptive problem has an optimal solution, then there will exist a set of weights for the corresponding nonpreemptive problem such that any optimal solution to the nonpreemptive problem is optimal to the preemptive problem. Further, any extreme point optimal solution to the preemptive problem is optimal to the nonpreemptive problem. Simply stated, there is always a set of weights that will produce the same extreme point solution(s) as the preemptive solution. However, there can be sets of weights such that the corresponding solution can not be obtained by any preemptive formulation. Thus, estimating relative weights is worth the trouble. Multiattribute value theory can be helpful in making this specification.

2.5   MULTIATTRIBUTE VALUE THEORY

As mentioned in the previous chapter, there is a multiattribute utility or value maximization problem underlying every approach to multiobjective optimization. In this case, it is a multiattribute value maximization problem since it is assumed that the objective functions, Z_h, are specified with certainty. To extract the functional form of the multiattribute value function implied by the goal programming model, it is useful to view the model in a slightly different form.

The first goal constraint

$$Z_1 - n_1 + p_1 = b_1$$

can be rewritten as

$$p_1 = b_1 - Z_1 + n_1.$$

Then, in the range where $p_1 > 0$, $Z_1 < b_1$ and $n_1 = 0$, the objective function term $w_1 p_1$ can be replaced by $w_1(b_1 - Z_1)$. The second goal constraint

$$Z_2 - n_2 + p_2 = b_2$$

can be rewritten as

$$p_2 = b_2 - Z_2 + n_2.$$

In the range where $p_2 > 0$, $Z_2 < b_2$ and $n_2 = 0$, the second objective function term $w_2 p_2$ can be replaced by $w_2(b_2 - Z_2)$. And, the third goal constraint

$$Z_3 - n_3 + p_3 = b_{3L}$$

can be rewritten as

$$p_3 = b_{3L} - Z_3 + n_3.$$

In the range where $p_3 > 0$, $Z_3 < b_{3L}$ and $n_3 = 0$, the third objective function term $w_3 p_3$ can be replaced by $w_3(b_{3L} - Z_3)$. Finally, the fourth goal constraint

$$Z_3 - n_4 + p_4 = b_{3R}$$

can be rewritten as

$$n_4 = Z_3 + p_4 - b_{3R}.$$

In the range where $n_4 > 0$, $Z_3 > b_{3R}$ and $p_4 = 0$, the fourth objective function term $w_4 n_4$ can be replaced by $w_4(Z_3 - b_{3R})$. The original goal programming objective function

Minimize $w_1p_1 + w_2p_2 + w_3p_3 + w_4n_4$

is equivalent to

Minimize $w_1(b_1-Z_1) + w_2(b_2-Z_2)$
            $+ w_3(b_{3L}-Z_3) + w_4(Z_3-b_{3R})$

or

Maximize $w_1Z_1+w_2Z_2+w_3Z_3+w_4(-Z_3)+$Constant.

The goal programming objective function can now be viewed as a specification of an additive value function. Therefore, the assumption of mutual preference independence defined in the previous chapter must hold for this model to be appropriate. Notice that the goal programming model is useful for some nonmonotonic preferences. The goal constraints are used to define the ranges where preferences are constant, monotonically increasing or monotonically decreasing. A separate term in the objective function is then used to model each range. If, however, preference for levels of an objective are monotonically increasing or monotonically decreasing over all feasible solutions the goal constraints are unnecessary. As the reformulated objective above shows, the monotonic objective, Z_1, can be optimized directly. Further, if preferences for each of the objectives are strictly monotonic it is easier to optimize a weighted combination of the objectives then to optimize a weighted sum of deviations because it is not necessary to specify aspiration levels.

There is an additional benefit to directly optimizing a weighted combination of the objectives. Changkong and Haimes (1983) and others have shown that optimizing a weighted linear combination of the objectives, where all the weights are nonnegative, will always yield a nondominated solution. Thus, if preferences for each of the objectives are monotonically increasing or monotonically decreasing, the objectives should be weighted, added and the

combination optimized directly. But how should the weights be
specified?

2.6   SPECIFYING THE WEIGHTS IN AN ADDITIVE
VALUE FUNCTION

The literature on multiattribute value theory describes a
number of techniques for specifying weights in the additive model.
These include multiple regression [Green and Rao, (1971) and
Beckwith and Lehman, (1973)]; Logit [Chapman and Staelin, (1982)];
multidimensional scaling [Carroll, (1972)]; mathematical
programming [Srinivasan and Shocker, (1973) and Horsky and Rao,
(1984)]; the analytic hierarchy process [Saaty, (1980)]; rating and
ranking of attributes [Edwards, (1977) and Einhorn and McCoach,
(1977)]; direct allocation of weights [Shoemaker and Waid, (1982)];
and equal weighting of standardized attributes [Einhorn and Hogarth,
(1975)]. Two of the more commonly used approaches, the rating and
ranking technique of Edwards (1977) and the linear programming
method of Horsky and Rao (1984) will be described below. These
methods illustrate the types of information required to estimate
weights in multiattribute value functions and the diversity of the
available procedures.

2.6.1 The SMART Procedure

The Simple MultiAttribute Rating Technique (SMART) was
developed by Edwards (1977) as a procedure for approximating
multiattribute value functions. The first step in SMART is to rank
the attributes by order of importance. Once this is done, each

attribute is rated as to its importance. This is accomplished by first assigning the least important attribute a weight of ten. For every other attribute the decision maker is asked, "How much more important (if at all) is it than the least important?" A number is then assigned which reflects this ratio. Thus, if attribute A is assigned a weight of ninety and attribute B is assigned a weight of thirty it means that attribute A is three times more important than attribute B. In performing this step, the decision maker is allowed to backup and reassess previously assigned weights until a consistent set of weights is obtained. These weights are then standardized by dividing each weight by the sum of all of the weights.

One criticism of this approach is that the attribute scales are not considered explicitly. It seems reasonable that the specific ranges of the attributes should impact the weights assigned to each attribute. However, direct judgments of importance may be insensitive to the ranges of the attributes under study. One way to address this problem is to use swing weights.

In the most common application of swing weights, the decision maker constructs two hypothetical vectors. The first vector has each attribute at its most preferred level; the second has each attribute at its least preferred level. The decision maker is then asked to assume that all attributes are currently at their worst level but one attribute can be improved to its best level. The question then is, "Which attribute would you change first? second? etc." This process establishes the rank order of the weights. To obtain weights from the rank orders, the change in value obtained when the first ranked attribute is improved from worst to best (while all other attributes remain at their worst level) is arbitrarily set to 100. The decision maker is then asked to express as a percentage of 100 the improvement in value obtained when each of the other attributes are individually improved from worst to best (again while all other attributes remain at their worst levels). Finally, weights are computed by standardizing these percentages.

Once weights are obtained, the next step in SMART is to scale the single-attribute value functions so that the maximum feasible attribute level is assigned a value of one and the minimum feasible attribute level is assigned a value of zero. Finally, the multiattribute value function is the sum of the scaled single-attribute value functions multiplied by the standardized weights.

In more recent applications of SMART [von Winterfeldt and Edwards, (1986)] strength of preference information has been incorporated in the development of single attribute values. Additionally, sensitivity analysis plays a greater role. The next section will describe a procedure for incorporating strength of preference information. Sensitivity analysis will be discussed later.

2.6.2 Mathematical Programming Procedures

Srinivasan and Shocker (1973) have developed a procedure that was generalized by Horsky and Rao (1984) which uses linear programming to determine attribute weights in a multiattribute value function. In this method the decision maker is first asked to identify which of two multiattribute outcomes is preferred. Suppose, for example, in a multiple-objective optimization setting, solution vector Z_1 is preferred to solution vector Z_2. Then, the numerical value associated with Z_1 should be greater than the numerical value associated with Z_2, $V(Z_1) > V(Z_2)$. Strength of preference information can be elicited by asking the decision maker to classify his or her preference as weak, moderate or strong. Suppose, the decision maker expresses a strong preference for solution vector Z_1 over solution vector Z_2 and a moderate preference for solution vector Z_2 over solution vector Z_3. Then, the difference in the numerical values associated with Z_1 and Z_2 should be greater than the difference in the numerical values associated with Z_2 and Z_3, $V(Z_1) - V(Z_2) > V(Z_2) - V(Z_3)$. The objective in this approach is to determine a set of weights that will minimize the inconsistencies in the stated preferences.

This problem can be formulated as a linear program. As an example, the preference statements given above would be represented as follows

Minimize: $\quad n_1 + n_2$

Subject to: $\quad V(Z_1) - V(Z_2) + n_1 - p_1 = 0$

$$[V(Z_1)-V(Z_2)] - [V(Z_2)-V(Z_3)]$$
$$+ n_2 - p_2 = 0$$
$$V(Z_a) \geq 0; \ a = 1, 2, 3$$
$$n_j \geq 0; \ j = 1, 2$$

where

$$V(Z_a) = \sum_{h=1}^{k} w_h Z_{ah}.$$

For the value function to be completely consistent with the preference information all of the negative deviations, n_j, should be zero. Thus, the objective is to minimize the sum of the negative deviations and obtain a value function as nearly consistent as possible. In general, there will be one constraint for each preference and strength of preference statement, and there will be one objective function term for each of these constraints. An additional constraint forcing the weights to sum to one is usually added.

Using this approach, standard linear programming sensitivity analysis can be used to analyze inconsistencies in the decision maker's preference information. Other types on sensitivity analysis will be discussed in the next section.

2.7   SENSITIVITY ANALYSIS

One advantage of the linear goal programming model is that standard linear programming procedures can be employed to solve the problem. This implies that the array of sensitivity analyses which are an integral part of the solution of linear programs can be applied to the goal programming model as well. In particular, shadow prices and range analysis can be used to evaluate the sensitivity of the optimal solution to changes in goal aspiration levels. Range analysis can also be used to investigate the effect on the optimal solution of changes in the weights in the objective function.

The usual linear programming sensitivity analysis is applied to one parameter at a time assuming that all other parameters are held constant. This assumption may not be appropriate when analyzing the weights in the goal programming objective function. Here changes in one weight will likely impact the other weights since they are relative weights and because they are frequently standardized. Barron and Schmidt (1988) have developed an approach which is useful for examining changes in the weights in an additive multiattribute value function.

The purpose of their procedure is to compute, for two nondominated multiattribute outcomes, the set of weights $\{w_{ah} \ h=1, 2, ..., k\}$ that make the numerical value associated with outcome Z_a exceed the numerical value associated with outcome Z_b by a specified amount δ. The weights are chosen so that they are as "close" as possible to the set of weights $\{w_{bh}\ h = 1, 2, ..., k\}$ which yielded Z_b. "Close" is measured by squared deviation. Thus, if a similar ("close") set of weights yields a different nondominated solution, the solution is sensitive to the choice of weights.

This problem is formulated as:

Minimize:
$$\sum_{h=1}^{k} (w_{ah} - w_{bh})^2$$

Subject to:
$$\sum_{h=1}^{k} w_{ah} Z_{ah} - \sum_{h=1}^{k} w_{ah} Z_{bh} = \delta$$

$$\sum_{h=1}^{k} w_{ah} = 1$$

$$w_{ah} \geq 0; \quad h = 1, 2, ..., k.$$

The solution to this problem, using Lagrange multipliers and ignoring the nonnegativity restriction is:

$$w_{ah} = \frac{(\delta - \sum_{h=1}^{k} D_h w_{bh})(k D_h - \sum_{h=1}^{k} D_h)}{k \sum_{h=1}^{k} D_h^2 - (\sum_{h=1}^{k} D_h)^2} + w_{bh}$$

where $D_h = Z_{ah} - Z_{bh}$. If this solution yields any negative weights w_{ah}, the nonnegativity restriction is violated and the problem must be solved as a quadratic programming problem.

Although Barron and Schmidt use squared deviation to measure the "closeness" of two sets of weights, other measures of distance may also be reasonable. For example the sum of the absolute deviations could be minimized by adding the following constraints to those in the Barron and Schmidt formulation

$$w_{ah} - n_h + p_h = w_{bh}; \quad h = 1, 2, ..., k$$

and minimizing the objective

Minimize: $\sum_{h=1}^{k} n_h + p_h.$

It is also possible to minimize the maximum deviation by adding the constraints

$$w_{ah} \cdot n_h + p_h = w_{bh}; \quad h = 1, 2, ..., k$$

$$n_h + p_h \leq D; \quad h = 1, 2, ..., k$$

to those in the Barron and Schmidt formulation and then minimizing the variable D. These two formulations have the advantage of being purely linear programming problems.

When this procedure is applied in a goal programming setting, Z_a can be any nondominated solution. Z_b is the optimal solution for the goal program with weights elicited from the decision maker. The weights w_{ah} will be those weights closest to the weights determined from the decision maker's preferences that will yield Z_a as the optimal solution. Thus, it is possible to see how small a change in the set of weights will yield a new solution. This contrasts with the sensitivity analysis which usually accompanies linear programming in that it considers the whole set of weights rather than one weight at a time.

In this section the closeness of two sets of weights is described in terms similar to those used in describing forecast error (i.e. squared deviation, absolute deviation and maximum deviation). This discussion could just as easily have been in terms of distance measures. The use of distance measures in decision making will be discussed in greater detail in Section 4.1.

The availability of well known clearly defined sensitivity analyses is a significant advantage of the linear goal programming model. These sensitivity analyses, however, do not relax the very restrictive assumptions of the model. As with linear programming models in general, the linear goal programming model assumes that all of the functional relationships (i.e. objectives and constraints) can realistically be modeled as linear functions.

Because the linear goal programming model is a multiple-objective model it requires additional assumptions regarding the decision process employed by the decision maker. As was discussed earlier in this chapter, the linear goal programming model implies mutual preference independence. This assumption cannot be relaxed by sensitivity analysis. The next chapter will present some approaches for generalizing the goal programming model and relaxing the physical and behavioral assumptions of the model.

2.8   REFERENCES

Barron, H. and C. P. Schmidt, "Sensitivity Analysis of Additive Multiattribute Value Models," *Operations Research*, 36 (1988), 122-127.

Beckwith, N. E. and D. R. Lehman, "The Importance of Differential Weights in Multiattribute Models of Consumer Attitude," *Journal of Marketing Research*, 10 (1973), 141-145.

Carroll, J. D., "Individual Differences and Multidimensional Scaling," *Multidimensional Scaling: Theory and Applications in Behavioral Sciences, Vol. I.: Theory*, Seminar Press, New York, (1972), 105-155.

Chapman, R. G. and R. Staelin, "Exploiting Rank Order Choice Set Data Within the Stochastic Utility Model," *Journal of Marketing Research*, 19 (1982), 288-301.

Edwards, W., "How to Use Multiattribute Utility Measurement for Social Decision Making, *IEEE Transactions on Systems, Man and Cybernetics*, SMC-7 (1977), 326-340.

Einhorn, H. J. and R. M. Hogarth, "Unit Weighting Schemes for Decision Making," *Organization Behavior and Human Performance*, 13 (1975), 171-192.

Einhorn, H. J. and W. McCoach, "A Simple Multiattribute Utility Procedure for Evaluation," *Behavioral Science*, 22 (1977), 270-282.

Green, P. E. and V. R. Rao, "Conjoint Measurement for Qualifying Judgmental Data," *Journal of Marketing Research*, 8 (1971), 355-363.

Horsky, D. and R. M. Rao, "Estimation of Attribute Weights from Preference Comparisons, *Management Science*, 30 (1984), 801-822.

Saaty, T. L., *The Analytic Hierarchy Process*, McGraw-Hill, New York, 1980.

Sherali, H. D., "Equivalent Weights for Lexicographic Multi-Objective Programs: Characterizations and Computations," *European Journal of Operational Research*, 39 (1982), 367-379.

Sherali, H. D. and A. L. Soyster, "Preemptive and Non-Preemptive Multi-Objective Programming: Relationship, Characterizations and Counter-Examples," *Journal of Optimization Theory and Applications*, 39 (1983), 173-186.

Shoemaker, P. J. H. and C. C. Waid, "An Experimental Comparison of Different Approaches to Determining Weights in Additive Utility Models," *Management Science*, 28 (1982), 182-196.

Srinivasan, V. and A. D. Shocker, "Linear Programming Techniques for Multi-Dimensional Analysis of Preferences," *Psychometrika*, 38 (1973), 337-369.

von Winterfeldt, D. and W. Edwards, *Decision Analysis and Behavioral Research*, Cambridge University Press, Cambridge, England, 1986.

2.9 OTHER RELEVANT READINGS

Dyer, J. S., "On the Relationship Between Goal Programming and Multiattribute Utility Theory," Discussion Paper No. 69, Management Science Study Center, Graduate School of Management, University of California at Los Angeles, Los Angeles, California, 1977.

Hannan, E. L., "An Assessment of Some Criticisms of Goal Programming," *Computers and Operations Research*, 12 (1985), 525-541.

Ignizio, J. P., *Goal Programming and Extensions*, D. C. Heath, Lexington, Massachusetts, 1976.

Ignizio, J. P., *Linear Programming in Single and Multiple Objective Systems*, Prentice-Hall, England Cliffs, New Jersey, 1982.

Ignizio, J. P., "Generalized Goal Programming: An Overview," *Computers and Operations Research*, 10 (1983), 277-289.

Lee, S. M., *Goal Programming for Decision Analysis*, Auerbach, Philadelphia, 1972.

Schneiderjans, M. J., *Linear Goal Programming*, Petrocelli Books, Princeton, New Jersey, 1984.

GENERALIZING GOAL PROGRAMMING

3.1 LINEAR GOAL PROGRAMMING

The linear goal programming model assumes that the decision maker's preference structure can be decomposed into an additive multiattribute value function. The model further assumes that each of the conditional single attribute value functions are linear or that they can each be adequately approximated by a linear function. This is shown graphically in Figure 3.1. As Figure 3.1 illustrates, the linear goal programming model uses a piecewise linear approximation for each of the single attribute value functions. The linear goal programming model uses one linear segment for each monotonically increasing or monotonically decreasing segment of the single attribute value function. This is the simplest piecewise linear approximation for these functions but it is also the least accurate. This chapter will discuss a procedure for obtaining more accurate piecewise linear approximations for the conditional single attribute value functions. In addition, the goal programming model will be generalized to accommodate multiplicative multiattribute value functions, which no longer require mutual preference independence.

3.2 PIECEWISE LINEAR APPROXIMATIONS OF SINGLE ATTRIBUTE VALUE FUNCTIONS

Keeney and Raiffa (1976) have shown how the characteristics of a decision maker's preference structure can be used to restrict the form of the decision maker's utility function. Their approach is

Figure 3.1

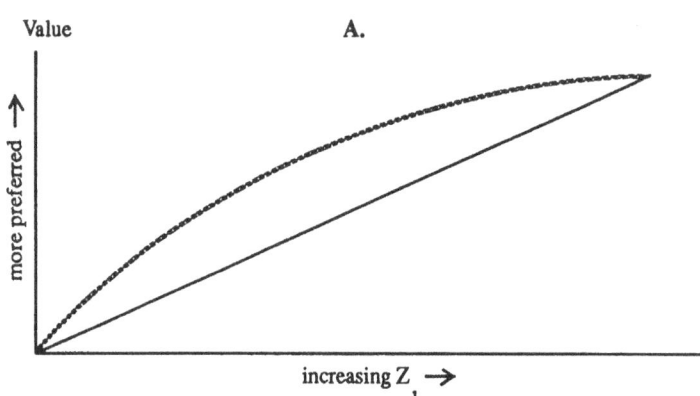

A.

Value

more preferred →

increasing Z_1 →

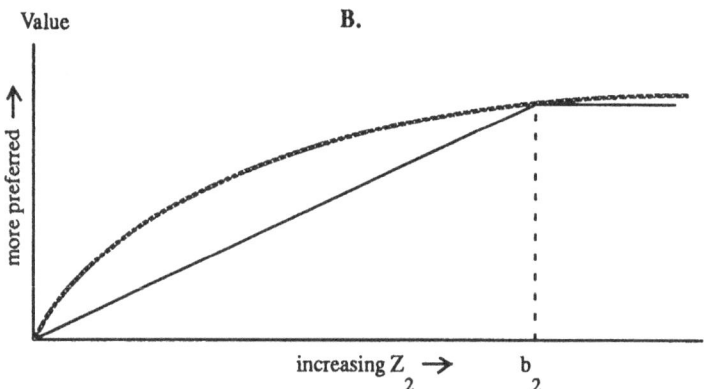

B.

Value

more preferred →

increasing Z_2 → b_2

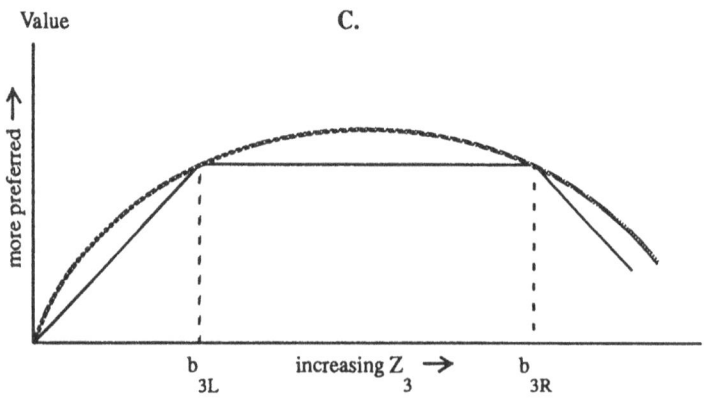

C.

Value

more preferred →

b_{3L} increasing Z_3 → b_{3R}

equally applicable to value functions and can be adapted for assessing linear piecewise approximations to nonlinear value functions.

To illustrate, suppose it is necessary to assess a value function $V(Z)$ for attribute Z. Further, suppose that the decision maker has indicated that his/her preferences are monotonically increasing in Z and that the marginal value of units of Z is decreasing. We can then choose two feasible numerical amounts for Z, Z_1 and Z_2, where $Z_2 > Z_1$ and we can arbitrarily assign $V(Z_1)$ and $V(Z_2)$ subject to the restriction that $V(Z_2) > V(Z_1)$. The points $[Z_1, V(Z_1)]$ and $[Z_2, V(Z_2)]$ can then be plotted on a graph as in Figure 3.2A.

From Figure 3.2A we can see that the decision maker's value function is restricted to the nonshaded area. Notice that for the value function to pass through the shaded area the function would necessarily be non-concave. But, since the decision maker stated that $V(Z)$ exhibits decreasing marginal value it cannot pass through the shaded area and it must be concave. Further, if we run a horizontal and a vertical line through the point $[Z_2, V(Z_2)]$ we can eliminate two other regions. The decision maker's value function cannot pass through the area above the horizontal line and to the left of the vertical line nor can it pass through the area below the horizontal line and to the right of the vertical line without violating monotonicity.

Now suppose we ask the decision maker to specify a point, Z_3, such that

$$V(Z_3) = \frac{V(Z_1) + V(Z_2)}{2}.$$

After plotting this point on the graph in Figure 3.2B, we can use the same logic as before to bound the decision maker's value function to the nonshaded region.

Figure 3.2

A.

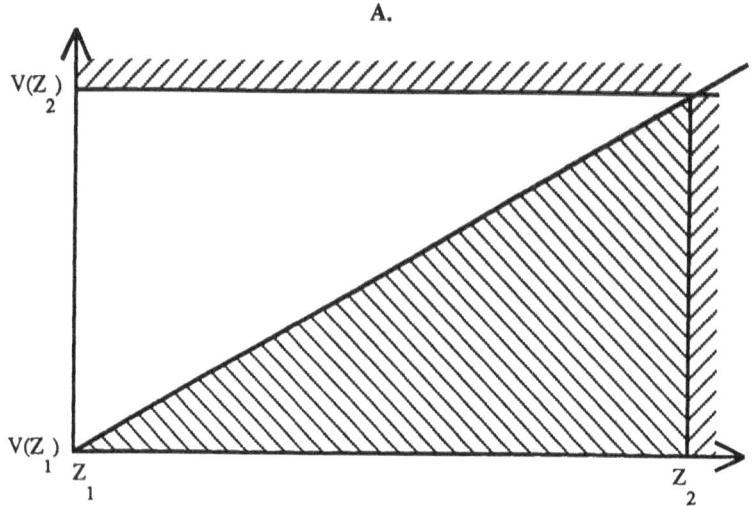

B.

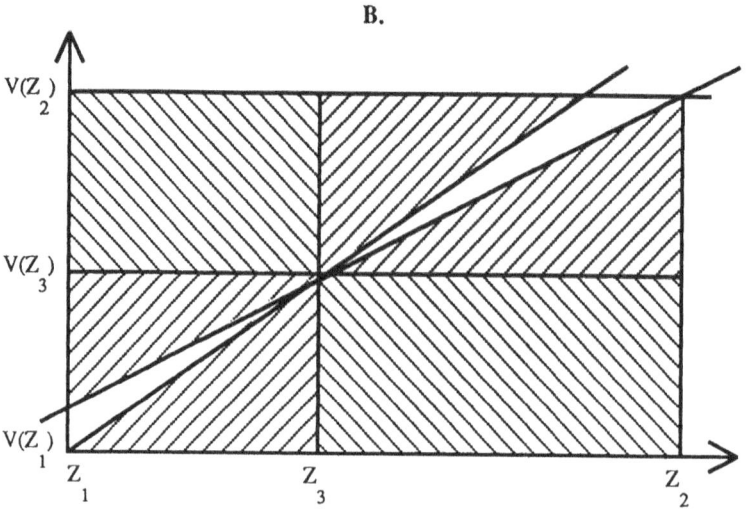

The lines which form these boundaries can also serve as a piecewise linear approximation to the true nonlinear value function. To illustrate, let $Z_1=0$, $Z_2=100$ and $Z_3=40$ also arbitrarily set $V(Z_1)=0$ and $V(Z_2)=1$ so that $V(Z_3)=0.5$. Then by geometric reasoning the true value function must lie between

$$V_1(Z) = \frac{Z}{80}$$

and

$$V_2(Z) = 0.167 + \frac{Z}{120}.$$

The appropriate linear pieces for maximizing $V(Z)$ would then be $V_1(Z)$ for $Z \leq 40$ and $V_2(Z)$ for $Z \geq 40$.

This approach also gives us a bound on the maximum error in our approximation. The bound will be the maximum vertical distance between $V_1(Z)$ and $V_2(Z)$. For a two piece approximation this will occur either at the origin or at the point where $V_1(Z)=1$. Figure 3.2B illustrates this for the example. At the origin, $V_1(Z=0)=0$ and $V_2(Z=0)=0.167$ so the maximum error at this point is $0.167-0=0.167$. At the point where $V_1(Z)=1$, $Z=80$ and $V_2(Z=80)=0.834$ so the maximum error at this point is $1-0.834=0.166$. Therefore, the maximum error in our estimate of $V(Z)$ over the range $0 \leq Z \leq 100$ is 0.167.

The analysis also gives us an indication of how this particular two piece approximation would compare with any other two piece linear approximation. If we were to assess the value of a point closer to (farther from) the origin, the error at the origin would tend to diminish (increase) while the error at $V_1(Z)=1$ would tend to increase (diminish). Thus, the best two piece linear approximation will have equal maximum error at these two points. This is very nearly the case for our example. If this were not the case or if the

maximum error were intolerably large, additional points could be
assessed and, applying the same logic as above, used to obtain closer
approximations with more linear segments.

   Once a satisfactory piecewise linear approximation has been
obtained for each conditional single attribute utility function, the goal
programming problem can be constructed.   The model for the
example would include the objective function terms

   Maximize: $w(Z_1/80) + w(Z_2/120) + ...$

(where w is the relative weight given to objective Z) and the
constraints

   Subject to: $Z = Z_1 + Z_2$

                      $Z_1 \leq 40$

                        .
                        .
                        .

The model would also include analogous objective function terms and
constraints for each of the other objectives as well as any additional
constraints in the problem. (Note: the first constraint above simply
sums the value of the linear segments and computes the optimal value
of objective Z.)

   At this point the linear goal programming model has been
generalized to allow for more precise approximations of the
conditional single attribute value functions. In the next section, the
model will be extended to include a multiplicative representation of
the multiattribute value function.

3.3   GOAL PROGRAMMING WITH A MULTIPLICATIVE VALUE FUNCTION

The linear goal programming model assumes that the decision maker's preferences can be modeled by an additive multiattribute value function of the form

$$V(Z) = \sum_{h=1}^{k} w_h\, V_h(Z_h)$$

where

$$0 \leq V_h(Z_h) \leq 1$$

and V_h is a conditional single attribute value function such that $V_h(Z_h^0) = 0$ and $V_h(Z_h^*) = 1$ for some Z_h^0 and Z_h^*;

$0 < w_h < 1$ are relative weights;

and

$$\sum_{h=1}^{k} w_h = 1.$$

A more general specification of the decision maker's preference structure would be the multiplicative model. The multiplicative model is of the form

$$1 + wV(Z) = \prod_{h=1}^{k} [1 + ww_h\, V_h(Z_h)]$$

where

$V_h(Z_h)$ are defined as before;

$0 < w_h < 1$ are scaling constants;

and

$-1 < w \leq \infty$ is a parameter.

This model is more general because it requires less restrictive assumptions about the decision maker's choice behavior. Specifically, the assumption of mutual preference independence is relaxed. The multiplicative model requires only preference independence. Recall that an attribute Z_h is preference independent of its complement $(Z_1, Z_2, ..., Z_{h-1}, Z_{h+1}, ..., Z_k)$ if the preference between two alternatives which differ only in attribute Z_h is dependent on the level of Z_h obtained with each alternative and is independent of the levels of its complementary attributes (all of these attribute levels are common to each alternative).

There are two disadvantages to this model in a goal programming setting. First, the multiplicative model as presented above is nonlinear even if all of the V_h are linear or piecewise linear. This problem, however, can be overcome as follows

The goal programming objective can be written as

Maximize: $V(\mathbf{Z})$

This is equivalent to

Maximize: $1 + wV(\mathbf{Z}) =$

$$\prod_{h=1}^{k}[1 + ww_h \, V_h(Z_h)], w > 0;$$

or

Minimize: $1 + wV(\mathbf{Z}) =$

$$\prod_{h=1}^{k}[1 + ww_h \, V_h(Z_h)],$$

$$-1 < w < 0;$$

which is in turn equivalent to

Maximize: $\log[1 + wV(\mathbf{Z})]$

$$= \log\{\prod_{h=1}^{k}[1 + ww_h \, V_h(Z_h)]\}$$

$$= \sum_{h=1}^{k} \log[1 + ww_h \, V_h(Z_h)], w > 0$$

or

Minimize:
$$\sum_{h=1}^{k} \log[1 + ww_h\, V_h(Z_h)],$$

$$-1 < w < 0.$$

This final expression is again a linear function as long as all V_h are linear or piecewise linear.

The second difficulty comes in assessing the scaling constants w_h and parameter w. Keeney and Raiffa (1976) describe the process for a multiplicative utility function. For a multiplicative value function, first scale the V_h and V as follows:

$$0 \le V_h(Z_h) \le 1,\; V_h(Z_h^*) = 1,\; V_h(Z_h^0) = 0,$$
$$0 \le V(Z)\; 1,\; V(Z^*) = 1 \text{ and } V(Z^0) = 0.$$

Then, the assessment of the w_h's is accomplished by asking the decision maker for a series of relative values or ratings. The necessary rating questions are of two basic types.

Rating Question 1. What is the relative value of an alternative $Z_h = (Z_1^0, ..., Z_{h-1}^0, Z_h^*, Z_{h+1}^0, ..., Z_k^0)$ as compared with Z^* and Z^0?

If we define the decision maker's answer as r_h then, due to the scaling imposed, $w_h = r_h$.

Rating Question 2. What level of Z_a in $Z_a = (Z_1^0, ..., Z_{a-1}^0, Z_a, Z_{a+1}^0, ..., Z_k^0)$ makes you indifferent between Z_a and $Z_b = (Z_1^0, ..., Z_{b-1}^0, Z_b, Z_{b+1}^0, ..., Z_k^0)$?

With the prescribed scaling, the value of these two alternatives can be equated yielding

$$w_a V_a(Z_a) = w_b V(Z_b).$$

Since at this point in the analysis both V_a and V_b have been assessed both $V_a(Z_a)$ and $V_b(Z_b)$ are easily found. Further, if $Z_b = Z_b^*$ the relationship is simply

$$w_a V_a(Z_a) = w_b.$$

A common practice in assessing the w_h's is to first rank the attributes, then to use Rating Question 1 to evaluate the largest w_h, and finally use Rating Question 2 to evaluate the magnitude of the other w_h's relative to the largest w_h. If the w_h's sum to one, the additive function is appropriate. If not, the multiattribute value function is multiplicative and the constant w must also be assessed.

The constant w can be found from the w_h values. We can evaluate $V(Z)$ at Z^* to get

$$1 + w = \prod_{h=1}^{k} (1 + w w_h).$$

If the sum of the w_h's is greater than one, the independence conditions will be satisfied if w is in the range $-1 < w < 0$ [see Keeney and Raiffa (1976)]. By iteratively evaluating the above expression, given the values of the w_h's, we can converge to the appropriate value of w. Note that if the right hand side of this expression is smaller than the left hand side the estimate of w is too large and vice versa.

If the sum of the w_h's is less than one , the independence conditions will be satisfied if w is positive. In this case, if the right hand side of the above expression is greater than the left hand side, the estimate of w is too large.

The above analysis extends the linear goal programming model to include better estimates of the conditional single attribute value functions and to include a multiplicative multiattribute value function. But, this analysis can be extended further to include some nonlinear goal programs.

3.4   NONLINEAR GOAL PROGRAMMING

The concepts described above for constructing a piecewise linear approximation of a nonlinear value function are consistent with separable programming. Separable programming is a nonlinear programming approach that can be applied to problems in which the objective function and constraints are separable. A function $f(\mathbf{x})$, $\mathbf{x} = (x_1, x_2, ..., x_n)$, is separable if it can be written as the sum of n single variable functions. Thus, both the additive and multiplicative value functions are separable and the goal programming model can be extended to include separable nonlinear objectives and constraints. This would be done by replacing each nonlinear objective or constraint with appropriate piecewise linear approximations and then applying the methods described in this chapter.

Monotropic optimization, developed by Rockafellar (1984), is a generalization of separable programming which includes both linear piecewise approximations and quadratic piecewise approximations as special cases. The extensions provided by monotropic programming might also be applied to further extend the goal programming model by improving on the approximations detailed above.

Ignizio (1976) describes two methods for solving nonlinear nonseparable goal programming models. One approach uses Taylor expansions to linearize the problem while the other applies pattern search. The use of search procedures in multiple-objective settings will be taken up in greater detail later in Chapters 5 and 6.

This chapter has described ways for generalizing the linear goal programming model.   In the next chapter, compromise programming, will be discussed. This discussion will establish the relationship between compromise programming, goal programming and multiattribute value theory.

3.5 REFERENCES

Ignizio, J. P., *Goal Programming and Extensions*, D. C. Heath, Lexington, Massachusetts, 1976.

Keeney R. L. and H. Raiffa, *Decisions with Multiple Objectives: Preferences and Value Tradeoffs*, John Wiley and Sons, New York, 1976.

Rockafellar, R. T., *Network Flows and Monotropic Optimization*, John Wiley and Sons, New York, 1984.

4
COMPROMISE PROGRAMMING

4.1 IDEAL SOLUTIONS

Yu (1973) and Zeleny (1974) define the ideal solution (Yu describes this solution as the "utopia point") as any solution that would simultaneously optimize each individual objective. In objective function space this point has the coordinates $Z(x^*) = [Z_1(x^*), ..., Z_k(x^*)]$, where x^* optimizes every $Z_h(x)$. It is an unusual case where there is a single solution which simultaneously optimizes all of the objectives. However, a representation of the unobtainable ideal solution can be obtained for any properly bounded set of alternatives by optimizing each $Z_h(x)$ separately. The coordinates of this representation are then given by $[Z_1(x_1^*), ..., Z_k(x_k^*)]$, where x_h^* optimizes the hth objective. Should any Z_h be unbounded, Z_h^* cannot be precisely specified. This case will be discussed in greater detail in Section 10.4. If the ideal solution is obtainable ($x_1^* = ... = x_k^*$) it is clearly the solution to the multiobjective linear program.

While the ideal solution point is not usually obtainable it is still of interest. It has played an important role in the works of Ackoff (1970, 1978). Zeleny (1975, 1976, 1977, 1982) argues that the ideal point is of primary importance in the human decision making process positing that, "To be as close as possible to the perceived ideal is assumed to be the rationale of human choice." Zeleny argues that managers make trade-offs among obtainable alternatives by comparing the distance of these alternatives from the ideal. The most preferred alternative is that which is (by some mental distance measure) closest to the ideal.

4.2   COMPROMISE FUNCTIONS

The objective of choosing a solution as close as possible to the ideal emphasizes the importance of distance and distance measurement in decision making. One of the most frequently used measures of distance in decision making is the family of L_p metrics, which can be described by the relationship

$$L_P = \{ \sum_{h=1}^{k} \pi_h [Z_h(x_h^*) - Z_h(x)]^P \}^{1/P}$$

where

x is a given solution

π_h is a gradation weight associated with the hth objective function, $\pi_h > 0$

P is the parameter that determines which of the family of L_p metrics is to be used, $1 \leq P \leq \infty$.

and, without loss of generality, we assume all objectives are maximization objectives.

In this measure, the effect of P is to place more or less emphasis on the relative contribution of individual deviations. The larger the value of P chosen, the greater is the emphasis given to the largest of the deviations forming the total. Ultimately, when $P=\infty$, the largest of the deviations completely dominates the distance measure. In addition to $P=\infty$, the values $P=1$ and $P=2$ are commonly used. $P=1$ implies the longest geometric distance between two points in that the deviations are simply summed over all dimensions. The L_1 metric is referred to as the "city block" or "Manhattan block" measure of distance. When $P=2$ we obtain the shortest geometric distance between two points, a straight line. Other values of P are not as easily interpreted. However, in this application the L_p metric is used to measure preference as the mental distance of any given solution from the ideal. Therefore, other values of P cannot be ruled out.

The π_h gradation weights are intended to reflect the relative importance of each objective and to modify the contribution to the total of each deviation. They can be estimated by procedures such as the simple multiattribute rating technique (SMART) described in Chapter 2.

In some cases, the value of the Lp metric can be influenced by the unit of measure chosen for a given attribute. For example, suppose the ideal solution has coordinates (5,3) in solution space and an available solution has coordinates (5,1). The equal weights L_2 metric is

$$L_2 = [(5-5)^2 + (3-1)^2]^{1/2} = 2.$$

Now suppose that the second attribute which was originally measured in meters is measured in centimeters yielding

$$L_2 = [(5-5)^2 + (300-100)^2]^{1/2} = 200.$$

In this second formulation, deviations in the first objective will almost certainly be overwhelmed by deviations in the second objective. This scale effect can be overcome by either of two alternative formulations

$$L_P = [\ \sum_{h=1}^{k} \pi_h \{[Z_h(x_h^*)-Z_h(x)]/Z_h(x_h^*)\}^P]^{1/P}$$

or

$$L_P = [\ \sum_{h=1}^{k} \pi_h \{[Z_h(x_h^*)-Z_h(x)]/[Z_h(x_h^*)-Z_h(x_{h*})]\}^P]^{1/P}$$

where in the second form x_{h*} is the solution which minimizes Z_h. In this second form each distance will be between 0 and 1.

4.3   COMPROMISE SOLUTIONS AND THE COMPROMISE SET

Given a weight vector, π, x^P is a compromise solution if and only if it minimizes Lp. The compromise solution has a number of desirable properties. First, if $\pi > 0$ it can be shown (see e.g. Chankong and Haimes, 1983) that x^P is always an efficient solution for any P in the range $1 \leq P < \infty$ and that at least one x^∞ is efficient.

The compromise set, X^C can also be defined for a given π as the set of all compromise solutions x^P, $1 \leq P \leq \infty$. Then, if we include only those x^∞ which are efficient, the compromise set will be a subset of the efficient solutions. For the linear case, the compromise set is completely determined if x^1 and x^∞ are known. In particular, for the two-objective case with $\pi > 0$, X^C is a set of efficient points which lie between x^1 and x^∞.

4.4   THE ANTI-IDEAL AND COMPROMISE PROGRAMMING

Zeleny (1982) discusses a concept which is a mirror image of the ideal solution. He terms this the anti-ideal. The question Zeleny poses is, "Do humans strive to be as close as possible to the ideal or as far away as possible from the anti-ideal?" His answer is that they probably do both. Thus, a compromise based on the anti-ideal may be equally appropriate.

A representation of the anti-ideal solution can be obtained for any properly bounded set of feasible alternatives. Again, without loss of generality, assuming that all objectives are maximization objectives we can obtain a representation of the anti-ideal solution by minimizing each $Z_h(x)$ separately. The coordinates of this representation are then given by $[Z_1(x_{1*}), ..., Z_k(x_{k*})]$, where x_{h*} minimizes the hth objective. A compromise function can then be formulated as

$$L_P = \{ \sum_{h=1}^{k} \pi_h [Z_h(x) - Z_h(x_{h*})]^P \}^{1/P}$$

and a compromise solution defined as the solution x^P which maximizes L_P for a given π. Compromise solutions based on the anti-ideal possess all of the same desirable properties as compromise solutions based on the ideal. However, since being as close as possible to the ideal is not equivalent to being as far away as possible from the anti-ideal the two compromise sets need not be the same. Although the compromise set is a subset of all efficient solutions each of these compromise sets may still contain many points. Thus, a decision maker may find it difficult to choose one solution. One way to further reduce the number of efficient solutions considered is to include only those solutions which are members of both compromise sets. Should a large number of solutions still remain, Zeleny (1974, 1975, 1982) describes an iterative method which will approach a single best compromise solution.

4.5   THE METHOD OF THE DISPLACED IDEAL

The ideal solution, $[Z_1(x_1^*), ..., Z_k(x_k^*)]$, depends on the set of available alternatives. After the set of alternatives is reduced to the compromise set, X^C, a new ideal solution can be constructed. The new point can be calculated from

$$Z_h^{*'} = \underset{x \in X^C}{\text{Max}} \, Z_h(x).$$

Using this new "displaced" ideal point, a new compromise set, $X^{C'}$ can be constructed analogously to the construction of X^C. Just as X^C is a subset of the set of all efficient solutions, $X^{C'}$ is a subset of X^C.

Once a reduced compromise set is obtained another new displaced ideal point can be constructed and the process repeated. By continuing the reduction of the compromise set using displaced ideal points, the compromise set should eventually enclose the new ideal point, thus terminating the process.

4.6   COMPROMISE PROGRAMMING, LINEAR GOAL PROGRAMMING, AND MULTIATTRIBUTE VALUE FUNCTIONS

The compromise programming objective function described in this chapter is directly related to the linear goal programming objective function discussed in Chapter 2. Further, like the linear goal programming objective function it implies a specific multiattribute value function.

Earlier the compromise objective function was defined as minimizing the L_P metric

$$\text{Min } L_P = \text{Min } \{ \sum_{h=1}^{k} \pi_h |Z_h(x_h^*) - Z_h(x)|^P \}^{1/P}$$

$$= \text{Min } \{ \pi_1 |Z_1(x_1^*) - Z_1(x)|^P + \ldots$$

$$+ \pi_k |Z_k(x_k^*) - Z_k(x)|^P \}^{1/P}$$

or equivalently for $1 \leq P < \infty$

$$\text{Min } L_P = \text{Min } \{ \pi_1 |Z_1(x_1^*) - Z_1(x)|^P + \ldots$$

$$+ \pi_k |Z_k(x_k^*) - Z_k(x)|^P \}.$$

The absolute values included in this formulation allow us to consider both maximization and minimization objectives.

The linear goal programming model was discussed in Chapter 2 and can be defined as follows

Minimize: $$\sum_{h=1}^{k} w_h |d_h|$$

Subject to: $$Z_h(x) + d_h = g_h; \quad h = 1,..,k$$

where d_h is the amount objective function Z_h deviates from aspiration level b_h and w_h is a weight used to measure the relative importance of deviations from aspiration level b_h. Then, by rewriting the constraints as

$$d_h = g_h - Z_h(x)$$

and substituting for d_h in the objective function we obtain

Minimize: $$\sum_{h=1}^{k} w_h |g_h - Z_h(x)|$$

$$= w_1 |g_1 - Z_1(x)| + \ldots$$

$$+ w_k |g_k - Z_k(x)|.$$

Finally, if we set $g_h = Z_h(x_h^*)$ the linear goal programming model is equivalent to compromise programming with $_h = w_h$ and $P = 1$.

In this light, we can view linear goal programming as a special case of compromise programming, [Yu (1985) also makes this connection] where we minimize the L_1 metric. Compromise programming is in one way less general than linear goal programming since it specifies a particular value for the goal aspiration levels, $g_h = Z_h(x_h^*)$. This specification, however, does seem quite reasonable.

This view of compromise programming also illuminates the relationship between multiattribute value functions and minimizing an Lp metric. The Lp metric used in compromise programming can be considered a particular specification of an additive value function. Thus, compromise programming, like linear goal programming, implies mutual preference independence.   Unlike linear goal programming, however, compromise programming does allow for nonlinear single attribute value functions through the specification of the P parameter.

4.7   REFERENCES

Ackoff, R. L., *A Concept of Corporate Planning*, Wiley, New York, 1970.

Ackoff, R. L., *The Art of Problem Solving*, Wiley, New York, 1978.

Chankong, V. and Y. Y. Haimes *Multiobjective Decision Making: Theory and Methodology*, North-Holland, New York, 1983.

Yu, P. L., "A Class of Solutions for Group Decision Problems," *Management Science*, 19 (1973), 936-946.

Yu, P. L., *Multiple-Criteria Decision Making*, Plenum Press, New York, 1985.

Zeleny, M., "A Concept of Compromise Solutions and the Method of the Displaced Ideal," *Computers and Operations Research*, 1 (1974), 479-496.

Zeleny, M., "The Theory of the Displaced Ideal," in M. Zeleny (Ed.) *Multiple Criteria Decision Making: Kyoto 1975*, Springer-Verlag, Berlin, 1975.

Zeleny, M., "The Attribute-Dynamic Attitude Model (ADAM)," *Management Science*, 23 (1976), 12-26.

Zeleny, M., "Adaptive Displacement of Preferences in Decision Making," in M. K. Starr, and M. Zeleny (Eds.) *Multiple Criteria Decision Making-TIMS Studies in Management Science*, North-Holland, Amsterdam, 1977.

Zeleny, M., *Multiple Criteria Decision Making*, McGraw-Hill, New York, 1982.

DECISION MAKING AND THE EFFICIENT SET

5.1 THE EFFICIENT SET

As was discussed in Section 1.2 the optimal solution to a multiple-objective optimization problem must be nondominated. But, there are likely to be a great many nondominated solutions for a given problem. The choice from among these nondominated solutions is determined by the decision maker's preferences among the multiple objectives. The goal and compromise programming approaches discussed can be used to specify an a priori functional representation of the decision maker's preference structure. This functional representation can then be optimized to obtain a single "best" nondominated solution. Goal programming and compromise programming are not the only methods which take this approach but, they do illustrate the general tactic.

An alternative approach is to generate the set of all efficient solutions (i.e. all nondominated solutions which are properly efficient) and to let the decision maker then choose an alternative from this set. Thus, the problem of modeling the inherent subjectivity in the problem is eliminated. The problem of generating the efficient set is well-defined mathematically and is a completely objective procedure. Generating the efficient set, however, does not necessarily enable the decision maker to identify the optimal solution. While eliminating the inferior points from consideration is important, the remaining efficient set is often so large that the decision maker is still left with a difficult choice.

The two objective case is special. The efficient set can usually be represented by a curve in two-space (i.e. a plot with $Z_1(\mathbf{x})$ on one axis and $Z_2(x)$ on the other axis). It can be generated by solving the parametric program

Maximize: $w_1 Z_1(x) + (1-w_1) Z_2(x)$

Subject to: $g(x) \leq 0$

                    $x \geq 0$

for $0 \leq w_1 \leq 1$, assuming that appropriate convexity conditions hold. Or, alternatively, it can be generated by solving the parametric program

Maximize: $Z_1(x)$

Subject to: $Z_2(x) \geq b_2$

                    $g(x) \leq 0$

                    $x \geq 0$

for $-\infty < b_2 < \infty$, regardless of the convexity conditions. Chankong and Haimes (1983) present the theoretical arguments which support these assertions.

For situations with more than two objectives most of the work applies to the linear case. Goicoechea, Hansen and Duckstein (1982), Changkong and Haimes (1983) and Steuer (1985) provide good surveys of the methods available. Many of these are direct extensions of the parametric programs given above. Most of this research has concentrated on generating the complete set of nondominated extreme points. However, in these problems the solution set is typically a nonconvex portion of the surface of the feasible region. Representing this solution set with just the nondominated extreme points may not be satisfactory. It is quite likely that the decision maker's true value function is nonlinear and that the most preferred solution is not an extreme point. In order to better represent the solution set, convex combinations of the nondominated extreme points can be generated to fill in areas of the solution set that are under-represented. The next section will present methods for generating convex combinations which result in a good representation of the complete solution set.

5.2   INTRA-SET POINT GENERATION

Steuer and Harris (1980) use the term intra-set point generation to refer to the process of generating a finite number of additional points from a continuous set given a subset of points already explicitly known. In this application the goal is to generate a set of points that are spread evenly over the entire nondominated solution set. Steuer and Harris recommend generating an over-abundance of points that can then be thinned down to a subset representative of the entire set. This thinning process will be discussed in the next section.

As was mentioned above, additional nondominated solutions can be identified by constructing convex combinations of the nondominated extreme point solutions. Steuer and Harris point out that the ability of this approach to efficiently represent the entire nondominated solution set depends on the probability distribution used to generate the weights used in these combinations. They have used two probability distributions, uniform and Weibull.

Steuer and Harris construct convex combinations of the form

$$\sum_{i=1}^{N} w_i Z_i$$

where

Z_i is the ith nondominated extreme point solution

w_i is the weight applied to the ith nondominated extreme point solution

N is the total number of nondominated extreme point
solutions

and

$$w_i = \frac{v_i}{\sum\limits_{i=1}^{N} v_i} \, .$$

The v_i are all drawn either from the (0-1) uniform distribution or the
Weibull distribution.   Steuer and Harris observed that for a
nondominated solution space which could be represented by a two
dimensional facet the convex combinations drawn from the uniform
distribution tend to cluster around the center of mass of the facet.
On the other hand, the convex combinations drawn from the Weibull
distribution tend to fill up the corners of the facet. Steuer and Harris,
therefore, recommend generating half of the extra points from the
uniform distribution and half from the Weibull distribution.   The
resulting nondominated solutions should provide a thorough coverage
of the entire nondominated solution space.

While generating additional nondominated solutions will
certainly provide a better representation of the nondominated
solution space, it will make the decision maker's choice of one best
solution even more difficult. The remaining sections of this chapter
will provide a discussion of several alternative decision aids which
may help with this choice.

5.3   FILTERING

Filtering is a term used by Steuer and Harris (1980) to refer
to a process in which subsets of points can be selected from a larger
finite set of points. Their idea is to discard the most "redundant"

points and to retain a subset of the most dissimilar points. To accomplish this a measure of the dissimilarity of pairs of points is used. The measure used is the family of L_p metrics which were discussed in Section 4.1. In this application, the L_p metric is described by the relationship

$$L_p = \{ \ \sum_{h=1}^{k} (\pi_h \, | \, Z_h^t - Z_h^i \, | \, ^P \}^{1/P} < d$$

where:

    k is the total number of objective functions

    π_h is a gradation weight associated with the hth objective function

    Z_h is the hth objective function

    t is the identification superscript of the nondominated solution point undergoing the dissimilarity test

    i is the identification superscript of the nondominated solution point currently retained by the filter

    P is the parameter that determines which of the family of L_p-metrics is to be used, $P \in \{1, 2, ..., \infty\}$

    d is a distance parameter that regulates the filtering process.

To initialize the filtering process a value for d is selected and the first nondominated solution point is arbitrarily held by the filter as one of the subset of dissimilar points. The next point held by the filter is determined by processing each of the remaining points through the filtering relationship. Each point which has a weighted distance value less than d is discarded. The remaining point with the smallest weighted distance value is then held by the filter.

    The process continues by computing the weighted distance value between the second point held by the filter and each of the points not discarded in the first step. Any point which has a weighted distance value less than d is discarded. The remaining point with the smallest sum of the two weighted distance values is retained by the filter and the process continues.

It should be noted that the number of points ultimately held by the filter is determined by the value of d specified. Unfortunately, there is no way to determine the number of points that will be held when a particular value of d is used other than by experimentation.

The value of P chosen determines the specific way in which the distance between any two points is measured. Frequently used measures for P are 1, 2 and ∞. These three metrics are identical to those used for sensitivity analysis in Section 2.7.

The π_h gradation weights also deserve mention. They are used to standardize the ranges for the different objective function values over the set of nondominated solutions being filtered. The $_h$ are computed as

$$\pi_h = [R_h \sum_{j=1}^{k} 1/R_j]^{-1}$$

where

> R_h is the difference between the maximum and minimum values of the hth objective function in the set of nondominated solutions being filtered.

This process of filtering will reduce the set of nondominated solutions presented to the decision maker to a more manageable number. Steuer and Harris, however, suggest a further fine tuning of the process. Specifically, they recommend the following steps

1. Generate a representative subset of points from the nondominated solution space.

2. Filter this subset down to a number of points (usually 5 to 10) that is manageable by the decision maker.

3. Have the decision maker identify the most preferred solution
    from the filtered points.

4. Using the filtering relationship, rank order the points identified in
    step 1 based on their distance away from the point identified in
    step 3.

5. Present the point identified in step 3 and a small number of the
    nearest neighbors to the decision maker for a final choice.

Thus, through interaction with the analyst the decision maker is able
to fine tune his/her decision.

5.4   CLUSTERING

An alternative to the filtering procedure of Steuer and Harris
has been suggested by Morse (1980). Morse has applied cluster
analysis to the problem of reducing the number of nondominated
solutions. Cluster analysis can be described as the process of forming
m groups or clusters of similar objects directly from a set of n objects.
Clusters are usually chosen to be mutually exclusive and in some way,
either qualitatively or quantitatively, the objects within a cluster are
similar.

Many algorithms have been developed for performing cluster
analysis. Morse tested a number of these in a multiobjective linear
programming setting.   He recommends a hierarchical clustering
procedure developed by Ward (1963). At each level hierarchical
clustering proceeds by joining two clusters to form successively larger
clusters. In Ward's method the distance of each point to the centroid
of its cluster is recorded. Distance could again be measured by any
L_p metric, but the L_2 metric is most commonly used. These distances
are then added up for each cluster and are called the error sum of

squares. At each step, the next two clusters to be joined are those for which the error sum of squares is minimized. Ward's method results in minimum variance spherical clusters. The clusters tend to be of fairly equal size and shape, thus dense regions of the solution space will not be under represented and sparse regions will not be over represented.

Once a manageable number of clusters has been formed from the complete set of nondominated solutions, the decision maker can be presented with a representative point from each cluster. This point may be the centroid of the cluster or any other element. The decision maker can then choose the most preferred of these points. Finally, the decision maker should be encouraged to examine a neighborhood around this point and to compare it with several other neighborhoods of points.

A similar clustering approach has been presented by Torn (1980). He samples points from the feasible region and then uses optimization techniques to obtain nondominated solutions. Finally, he proposes applying cluster analysis before any solutions are presented to the decision maker.

5.5   MATCHING AND GROUPING

Bard and Wambsganns (1990) have developed a matching based procedure which requires only pairwise comparisons of nondominated solutions. Their approach builds on three distinct mathematical concepts: 1) the use of convex cones to prevent the decision maker from making unnecessary decisions [Korhonen, et al. (1984)], 2) clustering of solutions into similar groups and elimination of all members of an inferior group, and 3) assessment of the decision maker's value function. Here the value function is used to provide a means of reducing errors inherent in the clustering procedure, rather than to perform rankings.

5.5.1 Algorithmic Approach

First the m alternatives are filtered to eliminate dominated solutions. Based on the size of the resulting nondominated set and the analyst's perception of limitations that may exist regarding the decision maker's involvement, a grouping strategy is selected. This grouping strategy specifies an upper limit on the number of clusters to be created given the current number, and the maximum number of elements permitted in the final group. The alternatives are then clustered in concert with the grouping strategy. Here, the decision maker is required to make pairwise comparisons between the centroids of the various groups until only one group remains with no more than a predetermined number of elements [cf. Steuer and Harris (1980)]. Clusters whose centroids fall within the bounds of any preference cones generated from previous comparisons are automatically eliminated from consideration.

When the final set of alternatives is obtained, the process enters the value function assessment stage. At this point, the decision maker's preferences, as reflected in the pairwise comparisons, are used to construct a linear program derived from the preference cones. The solution of this linear program yields an approximation of the decision maker's true value function. Using this approximate value function, all the alternatives previously discarded during clustering are reevaluated, and the "K best" are flagged for the decision maker to reexamine. If no solutions are flagged, the process terminates.

Otherwise, the solutions that have been identified for reexamination
are combined with the solutions in the "best group" and the clustering
is repeated. The process continues for a fixed number of iterations.
The alternatives remaining are offered to the decision maker for final
selection.

5.5.2 Clustering Alternatives

Let m be the number of nondominated solutions and k the
number of objectives. The basic goal is to cluster together similar
solutions so that the members of any one cluster differ from one
another as little as possible.

The clustering is done with a heuristic matching procedure in
which a weighted distance measure is used to represent the cost of
matching any two alternatives. Mathematically, the problem can be
described as follows. First, define a graph $G = (V,E)$ where V is the
set of all nodes and E is the set of all edges, and let $e_{ij} \in E$ represent
the edge which connects nodes $i,j \in V$. A subset of edges M is called a
matching if no two edges of M are incident to the same node. The set
of all matchings of G is denoted by U. Given edge weights, c_{ij}, the
cost of a matching, $C(M)$, can be computed by

$$C(M) = \sum_{e_{ij} \in M} c_{ij}.$$

The problem then is to minimize $[C(M): M \in U]$. Bard and
Wambsganss chose the L_2 metric weighted by the inverse of the
objective variance for computing the edge weights. Specifically, the
edge weight between alternatives Z^i and Z^j is defined as

$$c_{ij} = [\sum_{h=1}^{k} \{ w_h [Z_h^i - Z_h^j]^2 \}]^{1/2}$$

where

$$w_h = [\ \sum_{\alpha=1}^{m} (\ Z_h{}^\alpha - \bar{Z}_h\ )\ ]^{-1}$$

and

$$\bar{Z} = \max\ [Z_h{}^i\colon\ 1 = 1,2,...,m].$$

The matching procedure requires that the graph have an even number of nodes, so if m is odd, a dummy node (call it m+1) is added with $c_{i,m+1} = 0$, $i = 1,2,...,m$. This step generates a minimum cost matching, producing m/2 pairs of nodes or clusters. The weights are then redefined as the distance between the centroids of each cluster, and a second matching is performed. The procedure continues until the number of clusters remaining is less than or equal to some specified upper bound.

5.5.3 Grouping Strategies

A grouping strategy is used which specifies an upper limit on the number of clusters to be created given the current number, and the maximum number of elements permitted in the final group. If, for example, the upper limit is set at 10, the matching algorithm is applied sequentially until 10 or fewer clusters remain. The general form of the grouping strategy, S, is a vector of s paired values linking the number of clusters, n_L, with the number of alternatives remaining, b_L, (L = 1,2,...,s), followed by an upper limit on the number of elements permitted in the final group, m^*. Notationally, S = $[(n_1,b_1),...,(n_s,b_s),\ m^*]$. Given m clusters, S is interpreted as follows.

Form at most n_1 clusters if $m > b_1$;

form at most n_2 clusters if $b_1 \geq m > b_2$;

$$\vdots$$

form at most n_s clusters if $b_{s-1} \geq m \geq b_s$.

Thus, a grouping strategy represents a deliberate tradeoff between an expected number of decision maker queries and the probability of experiencing some degradation in quality or value of those alternatives arising in the final set.

Specifying S also has the effect of imposing an upper bound on the number of queries required of the decision maker. For example, consider a set of $m = 100$ alternatives. If $S = [(10,30), (4,10), (2,1), 4]$, and if the number of elements in each cluster were equal, clustering would continue until 10 groups, each containing 10 elements, had been generated. Nine pairwise comparisons would be required to reduce this set to a single cluster of ten alternatives. This set would in turn be clustered into 3 groups (the 10 elements would be clustered into 5 pairs, a dummy node would be added, and the 6 clusters would in turn be clustered into 3 groups, which is less than or equal to the upper bound of 4). Hence, two additional pairwise comparisons would be required to reduce the set of 10 alternatives to a single set of 2 to 4 alternatives. Summarizing, one would expect no more than $9 + 2 = 11$ total pairwise comparisons to reduce the 100 alternatives to a single set of no more than 4 alternatives.

5.5.4 Value Function Assessment

It is well known [Keeney (1977)] that the preference of one group's mean over that of another does not necessarily imply that the solution of greatest overall value lies within the group with the preferred mean. To account for the possibility that the best solution was discarded during the clustering procedure, Bard and Wambsganss approximate a value function based on the decision maker's stated preferences. They estimate an additive value function using a linear programming approach which allows for inconsistent responses from the decision maker. This value function is then used to identify additional solutions that should be reconsidered by the decision maker.

The decision maker's preferences are expressed as either preference or indifference denoted by (>-) and (~), respectively. Hence, if the decision maker stated that Z^i >- Z^j, this is denoted as $(i,j) \in \{>-\}$. If the decision maker stated that $Z^i \sim Z^j$, then $(i,j) \in \{\sim\}$. The value function can be approximated by exploiting a mathematical formulation that is guaranteed to possess a feasible solution.

$$\text{minimize} \quad \sum_{(i,j)\in\{>-\}} (\Phi_{ij}) + \sum_{(i,j)\in\{\sim\}} |\theta_{ij}|$$

subject to

$$\sum_{h=1}^{k} W_h[V_h(Z_h^i) - V_h(Z_h^j)] + \Phi_{ij} \geq \delta,$$

$$\text{all } (i,j) \in \{>-\}$$

$$\sum_{h=1}^{k} W_h[V_h(Z_h^i) - V_h(Z_h^j)] + \theta_{ij} = 0,$$

$$\text{all } (i,j) \in \{\sim\}$$

$$\sum_{h=1}^{k} W_h = 1,$$

$$W_h \geq \delta,$$

$$\Phi_{ij} \geq 0,$$

θ_{ij} unrestricted

where δ is an arbitrarily small positive number, and the value function, V_h, is given. Observe that this objective function implies that the L_1 metric is to be minimized.

Because this linear program always has a solution, it is always possible to produce an additive value function approximation to the decision maker's true value function based solely on pairwise comparisons. The accuracy of the fit depends upon the nature of the true value function, and the forms selected for V_h. Bard and Wambsganss present results for V_h taking both additive-linear and additive-quadratic forms.

5.6   SECTIONING

The problem of choosing the preferred nondominated solution from the nondominated solution set may also be viewed as a search for the point which optimizes the decision maker's value function. This type of problem has been treated extensively in the literature on nonlinear optimization or search techniques.

Friedman and Savage (1947) describe one of the simplest of the multivariable search techniques, called sectioning, in which one variable at a time is altered while holding all others constant. For a function, $f(x_1, x_2, ..., x_n) = f(\mathbf{x})$, the search proceeds by finding the value of x_j which optimizes $f(\mathbf{x})$ given that all x_i, $i \neq j$; are held constant. The process is repeated through all of the x's as many times as necessary until there is no longer any improvement in $f(\mathbf{x})$.

A variation of sectioning can be applied to the problem of choosing the most preferred nondominated solution. Given a set of nondominated solutions to choose from, the problem is one of finding a nondominated solution, $(Z_1, Z_2, ..., Z_k) = Z$, which maximizes the decision maker's value function $V(Z)$. The first distinction in this application of sectioning is that unlike $f(x)$, $V(Z)$ is not known explicitly. The $V(Z)$, however, is only needed to evaluate the preference between two Z's. The decision maker should be able to express this preference directly without explicitly assessing $V(Z)$.

The search can then proceed from an arbitrarily selected nondominated solution by asking the decision maker if $V(Z)$ could be improved by changing Z_h. If the answer is no, the question is repeated for Z_{h+1}. If the answer is yes, the Z's are rank ordered from best to worst by Z_h. Then the decision maker is asked to identify the most preferred of the higher ranked solutions. Continuing from this solution, h is incremented and the process repeated. These steps are cycled through as necessary until no better solutions are identified.

For example, consider the following set of nondominated solutions which are rank ordered by Z_1 (a maximization objective)

Rank	Z_1	Z_2	Z_3
1	48	32	-16
2	16	24	0
-> 3	16	0	16
4	5	21	5
5	0	8	16

Arbitrarily starting with solution (16, 0, 16) the decision maker would be asked to compare this solution to each of the higher ranked (by Z_1) solutions. Suppose the 2nd ranked solution is preferred but the 1st ranked is not. The solutions are reordered by Z_2 (also a maximization objective)

Rank	Z_1	Z_2	Z_3
1	48	32	-16
-> 2	16	24	0
3	5	21	5
4	0	8	16
5	16	0	16

and the search is continued with solution (16, 24, 0). This solution is still preferred to the 1st ranked solution, so the solutions are reordered by Z_3 (also a maximization objective) and the search continued.

Rank	Z_1	Z_2	Z_3
1	16	0	16
2	0	8	16
3	5	21	5
-> 4	16	24	0
5	48	32	-16

At this point, if none of the higher ranked solutions are preferred, the search is terminated. If a preferred solution is identified, the solutions would be reordered by Z_1 and the search would continue.

5.7   A STOCHASTIC SCREENING APPROACH

An assumption of both single and multiple objective linear programming is that the coefficients of the objective function(s) and constraints are known and constant. In many applications, however, this assumption is violated. Ringuest and Graves (1989) describe an application of multiobjective linear programming to the problem of selecting from among competing R&D projects. In this application, one of the objectives is a function of future cash flows that are anticipated from various R&D projects. Ringuest and Graves use the expected values of the random variables (annual cash flows) for the objective function coefficients. Thus, the risk associated with uncertain future cash flows is not explicitly considered. In fact, the implication is that the risk associated with each of the different R&D projects is approximately the same or in more general terms, the variances of the uncertain objective function coefficients are equal.

In some cases it may be possible to consider risk explicitly and to use this information to screen out solutions that are too risky. Goicoechea, Hansen and Duckstein (1982) describe a stochastic programming procedure that can be used for this purpose. In order to apply stochastic programming an approximate probability distribution must be known for each of the objective function coefficients. The following probability statement can then be formed

$$P[Z(x) \geq d] \geq 1-\alpha$$

where

$$Z(x) = c_1 x_1 + c_2 x_2 + \ldots + c_n x_n$$

α is an element of $R[0,1]$

d is a real number which specifies a desired level for $Z(x)$

c_i is a random variable, $i = 1, 2, \ldots, n$.

It is usually assumed that the c_i's are normally distributed, $N[E(c_i), \mathrm{var}(c_i)]$. With the normality assumption, the probability statement can be rewritten as

$$P \left[\frac{Z(x) - \sum_{i=1}^{n} E(c_i) x_i}{[x^t Bx]^{1/2}} \geq \frac{d - \sum_{i=1}^{n} E(c_i) x_i}{[x^t Bx]^{1/2}} \right] \geq 1-\alpha$$

where

B is a symmetric variance-covariance matrix.

This second probability statement is true if and only if

$$\frac{d - \sum_{i=1}^{n} E(c_i) x_i}{[x^t Bx]^{1/2}} \leq K_\alpha$$

where

K_α is a standard normal value such that $\Phi(K_\alpha) = \alpha$

Φ represents the cumulative distribution function for the standard normal distribution.

Finally, this statement can be rewritten as

$$\sum_{i=1}^{n} E(c_i)x_i + K_\alpha[x^t Bx]^{1/2} \geq d.$$

This last inequality can be used in two different ways. First, K_α can be specified and the inequality can be added to the math programming problem as a constraint. The solution to this new math program would yield an x, Z and the value of d which could be achieved with α probability. Obtaining this solution(s) might be quite difficult because the inequality is nonlinear. Note that one nonlinear constraint (each of which could specify a different K_α value) is added to the problem and a separate d_h is obtained for each objective function, Z_h, with uncertain coefficients.

A second approach is available which eliminates this nonlinearity. The multiobjective linear program and intra-set point generation can be used to construct a subset of the nondominated solutions. Then a value (or values) of d_h can be specified for each objective function, Z_h. Finally, for each nondominated solution, the inequalities can be solved for K_α to yield the probability that any particular solution will produce a value of at least d_h for objective function Z_h. These probabilities can be viewed as a measure of the risk associated with each nondominated solution. Any solution associated with too high a risk (too low a probability of obtaining d_h) can be eliminated from further consideration. The d_h's are similar to the aspiration levels in goal programming. However, the d_h's are not used in solving the multiobjective linear program so there is no chance of producing dominated solutions as in goal programming.

These stochastic inequalities are one way to explicitly consider risk in the analysis of a multiobjective problem. Other problems associated with risk and consequences which occur over time will be considered in Chapter 9.

5.8   REFERENCES

Bard, J. F. and M. Wambsganss, "An Interactive MCDM Procedure Using Clustering and Value Function Assessment, Operations Research Group, Department of Mechanical Engineering, University of Texas, Austin, Texas (1991).

Chankong, V. and Y. Y. Haimes, *Multiobjective Decision Making: Theory and Methodology*, North-Holland, New York, 1983.

Friedman, M. and L. S. Savage, "Planning Experiments Seeking Maxima," in *Selected Techniques of Statistical Analysis*, C. Wisenhart, M. W. Hastay and W. A. Wallis, Eds. McGraw-Hill, New York, 1947.

Goicoechea, A., D. R. Hansen and L. Duckstein, *Multiobjective Decision Analysis with Engineering and Business Applications*, John Wiley and Sons, New York, 1982.

Keeney, R. L., "The Art of Assessing Multiattribute Utility Functions," *Organizational Behavior and Human Performance*, 19, (1977), 267-310.

Korhonen, P., J. Wallenius and S. Zionts, "Solving the Discrete Multiple Criteria Problem Using Convex Cones," *Management Science*, 30 (1984), 1336-1345.

Morse, J. N., "Reducing the Size of the Nondominated Set: Pruning by Clustering," *Computers and Operations Research*, 7 (1980), 55-66.

Ringuest, J. L. and S. B. Graves, "The Linear Multi-Objective R&D Project Selection Problem," *IEEE Transactions on Engineering Management*, 36 (1989), 54-57.

Steuer, R. E. and F. W. Harris, "Intra-Set Point Generation and Filtering in Decision and Criterion Space," *Computers and Operations Research*, 7 (1980), 41-53.

Steuer, R. E., *Multiple Criteria Optimization: Theory, Computation, and Application*, John Wiley and Sons, New York, 1986.

Torn, A. A., "A Sampling-Search-Clustering Approach for Exploring the Feasible/Efficient Solutions of MCDM Problems. *Computers and Operations Research*, 7 (1980), 67-79.

Ward, Jr., J. E., "Hierarchical Grouping to Optimize an Objective Function," *Journal of the American Statistical Association*, 59 (1963), 236-244.

5.9   OTHER RELEVANT READINGS

Graves, S. B., J. L. Ringuest and J. F. Bard, "Recent Developments in Screening Methods for Nondominated Solutions in Multiobjective Optimization," *Computers and Operations Research* (forthcoming).

INTERACTIVE METHODS

6.1 THE GENERAL INTERACTIVE APPROACH

Some methods for reducing the set of nondominated solutions allow the decision maker to participate in the process of screening solutions to achieve a subset which can be evaluated directly. This participation begins only after a representative set of nondominated solutions has been generated. It is, however, possible to bring the decision maker into the solution process at an earlier stage. The class of methods which do this are called interactive methods or methods of progressive articulation of preferences.

These methods differ from the methods for reducing the set of nondominated solutions in that they do not begin by generating a full set of solutions representative of the nondominated solution space. Interactive methods in general begin by identifying one or more nondominated solutions. The decision maker is then queried about tradeoffs among these solutions or preferences among the solutions. The problem is then modified based on the decision maker's responses, a new solution is produced and new information is solicited. These steps are repeated iteratively until the decision maker is satisfied with the solution, provided a satisfactory solution exists.

Interactive methods may be classified by the type of information that is required from the decision maker. Some methods require explicit information regarding the tradeoff among the levels of objectives at each stage of the solution process. Others require the decision maker to indicate the acceptability of the objective function levels (i.e. identify the best or worst solution) at each stage.

Tradeoffs among objective function levels are then implied from these preferences. Some examples describing members of each of these classes of methods will follow. A detailed survey of interactive methods including specification of the algorithms can be found in either Goicoechea, Hansen and Duckstein (1982) or Steuer (1986).

6.2   EXAMPLES OF INTERACTIVE METHODS

Geoffrion, Dyer and Feinberg (1972) have developed an algorithm which is essentially an interactive application of the Frank-Wolfe search procedure (1956). This algorithm asks the decision maker to make pairwise comparisons between objective function vectors. Haimes and Hall (1974) proposed the surrogate worth tradeoff method. This method requires the decision maker to assess the relative value of the tradeoff of marginal increases and decreases between any two objectives. Zionts and Wallenius (1976, 1983) present a method in which the decision maker's assessment of pairwise comparisons between alternatives and tradeoffs are used to construct local linear approximations to the decision maker's value function.

Benayoun, et al. (1971), Steuer (1977), Steuer and Wood (1986) and Korhonen and Laakso (1986) have all developed interactive algorithms which require the decision maker to provide preference information. This type of information is generally easier for the decision maker to assess and requires fewer assumptions regarding the structure of the decision maker's underlying value function. For these reasons, two algorithms which describe the type of implicit tradeoff information which can be elicited from the decision maker will be described in greater detail.

6.3 SIMPLIFIED INTERACTIVE MULTIPLE OBJECTIVE LINEAR PROGRAMMING (SIMOLP)

Based on a review of earlier research, Reeves and Franz (1985) constructed a set of characteristics that they consider to be desirable properties of interactive algorithms. These properties are

1. A minimum of inputs should be required from the decision maker. This includes inputs used to construct weights and any other quantitative assessments.

2. The decision making process should be simplified. A manageable number of alternatives should be presented to the decision maker at each step and the decision maker should be asked to make relatively few choices from among these alternatives.

3. The decision maker may change his or her mind and the decision maker may gain additional insights into the problem during the interaction with the solution algorithm. Therefore, the algorithm should be able to backtrack to solutions presented at earlier iterations.

4. The algorithm should generate a solution in the neighborhood of the best solution after just a few iterations.

5. The decision maker should be allowed to continue using a familiar decision making process. Choices at each stage in the algorithm should be structured so that they are similar to choices which are commonly made without the algorithm.

6. The algorithm should be applicable to large, realistic problems.

The SIMOLP procedure was developed by Reeves and Franz with these six properties in mind. SIMOLP allows the decision maker to identify preferred nondominated solutions interactively with a reasonable amount of effort without generating a complete representation of the entire nondominated set. This procedure optimizes a single weighted linear combination of the multiple objectives at each stage of the algorithm. Thus, the multiobjective problem is solved through a sequence of single objective linear programming problems. The weights are determined with the aid of simple preference information elicited from the decision maker and a new linear approximation to the decision maker's value function is assessed and optimized at each iteration.

The Reeves and Franz procedure has some elements in common with the NonInferior Set Estimation (NISE) method of Cohon (1978, 1979). Both procedures identify nondominated solutions and construct hyperplanes which pass through them in objective function space. The NISE procedure, however, was developed to generate all nondominated solutions to problems restricted to two objective functions and is not interactive. SIMOLP is also similar to a Boundary Point Ranking procedure (BPR) of Hemming (1976). BPR makes use of the Nelder and Mead (1965) unconstrained nonlinear single objective search algorithm. It is an interactive algorithm but BPR is much more complicate than SIMOLP, requiring more complex information from the decision maker.

SIMOLP includes the following steps:

1. Solve the k single objective linear programming problems defined as

   Maximize: $Z_h(\mathbf{x})$; $h = 1, 2, ..., k$
   Subject to: $\mathbf{g(x)} \le \mathbf{0}$
                    $\mathbf{x} \ge \mathbf{0}$.

   to obtain k nondominated solutions, Z^h, $h = 1, 2, ..., k$; and the corresponding k efficient points, $\mathbf{x}^h$, $h = 1, 2, ..., k$. Let $N^* =$

$\{Z^h, h = 1, 2, ..., k\}$, $E^* = \{x^h, h = 1, 2, ..., k\}$ and have the decision maker review the elements of N^*. If the decision maker is satisfied with the most preferred element of N^* (i.e, the best available solution) the procedure terminates. If the decision maker wishes to try to improve on this solution, set $h = k$ [$Z_h(x)$; $h = 1, ..., k$ will always correspond with the original k objective functions in the problem] and continue with step 2.

2. Set $h = h + 1$. Form the hyperplane $Z_h(x)$ which passes through the k elements of N^* [$Z_h(x)$, $h > k$ will always correspond with a hyperplane]. Solve the linear programming problem defined as

$$\text{Maximize:} \qquad Z_h(x)$$
$$\text{Subject to:} \qquad g(x) \leq 0$$
$$x \geq 0.$$

to obtain the nondominated solution Z^h and the corresponding efficient point x^h.

3. a. If $Z^h \in N^*$ and if Z^h is preferred to at least one element of N^*, replace the least preferred element in N^* with Z^h and go to step 2.

    b. If $Z^h \in N^*$ or if the decision maker does not prefer Z^h to any element of N^* stop. The solution is the most preferred element of N^*.

In this algorithm, step 1 initializes the solution procedure. Steps 2 and 3 are then repeated iteratively until the decision maker identifies a preferred solution. At each intermediate iteration the decision maker is asked to identify the least preferred solution from N^*. At the termination of the algorithm the decision maker must identify the most preferred solution from N^*. In this algorithm a solution discarded at one iteration may be encountered again in a future iteration. Thus, the decision maker may change his or her mind during the solution process and reconsider a previous decision as in characteristic 3 of Reeves and Franz.

Two additional technical points should be mentioned. First, the hyperplane constructed in step 2 of each iteration is found by solving the $k \times k+1$ homogeneous system of linear equations

$$\sum_{h=1}^{k} w_h Z_h^j - w_{k+1} = 0, Z^j \in N^*.$$

Second, if any of the linear programs at step 1 or step 2 have alternate optimal solutions a possibility exists that the solution obtained will be dominated (for the full multiobjective linear program) by one or more of the alternate solutions. To ensure that only nondominated solutions are obtained in these problems Reeves and Franz suggest adding a term to the objective function of the form

$$\alpha \sum_{h=1}^{k} Z_h$$

where

α is a small positive constant.

SIMOLP is an interactive algorithm requiring relatively simple preference information from the decision maker which is applicable to linear problems. It is similar to some nonlinear search algorithms. In the next section an interactive algorithm which is applicable to more general nonlinear problems will be presented. This algorithm is also grounded in the search literature.

6.4   INTERACTIVE MULTIOBJECTIVE COMPLEX SEARCH

Ringuest and Gulledge (1985) have developed an interactive version of complex search that is applicable to multiobjective problems. Complex search was originally presented by Box (1965). It is a direct search procedure designed for single objective nonlinear programming problems with linear or nonlinear inequality constraints.

Direct search methods are defined as those search methods which do not require derivatives. These methods simply progress through a sequence of points according to some algorithm. As a general rule gradient and second derivative methods converge faster than direct search methods, particularly when solving unconstrained nonlinear programming problems. Direct search methods, however, are applicable to a broader range of problems, since they do not require regularity and continuity of the objective function and the existence of derivatives.

Spendley, Hext and Himsworth (1962) developed one of the earliest direct search algorithms in connection with the statistical design of experiments. In the search for a maximum of a single objective function $f(\mathbf{x})$, trial $\mathbf{x}$'s are selected at points located at the vertices of a polyhedron or simplex. The objective function is evaluated at each of the vertices of the simplex and a projection is made from the point yielding the lowest value for the objective function through the centroid of the simplex. The point yielding the smallest objective function value is deleted and a new simplex is formed composed of the remaining points from the old simplex and the one new point from the projection through the centroid. This process is continued with rules for reducing the size of the simplex and for preventing cycling near the optimum. It provides a derivative free search in which the step size is fixed at any stage but the search direction is variable.

This method does not provide for acceleration of the search and it encounters difficulty in curving valleys or ridges. Nelder and

Mead (1965) present an algorithm termed the "flexible polyhedron" method in which the simplex is allowed to alter in shape and thus no longer stays a simplex. This method maximizes a single function of n variables using $n+1$ vertices of a flexible polyhedron. The vertex which yields the smallest value of $f(\mathbf{x})$ is projected through the centroid of the remaining vertices. Improved values of the objective function are found by replacing the point with the smallest value of $f(\mathbf{x})$ by better points until the maximum of $f(\mathbf{x})$ is found.

The complex method of Box (1965) was designed for single objective nonlinear programming problems with linear or nonlinear inequality constraints. This procedure is a further modification of the simplex search and the flexible polyhedron search. A difficulty with the simplex and flexible polyhedron methods is that when the algorithm repetitively encounters a constraint it is necessary to withdraw the nonfeasible vertex until it becomes feasible. After many of these withdrawals the polyhedron collapses into n-1 or fewer dimensions and the search slows. Further, if the constraint ceases to be active, the collapsed polyhedron cannot easily expand back to the full n dimensions. Box avoided these difficulties by using a polyhedron with more than n vertices, termed a complex.

The complex method employs $n+1$ or more vertices, p, each of which must fall in the feasible region. An initial vertex $\mathbf{x}_1^{(0)}$ is chosen and random numbers are used to generate the remaining p-1 vertices. An objective function value is then obtained at each vertex and the vertex having the worst objective function value is replaced by a new one. The new vertex is projected along the line joining the point to be replaced and the centroid of the remaining points. It is placed at a distance at least as great as the distance from the point being replaced to the centroid. If a constraint is violated or if the new vertex gives the worst objective function value it is replaced by a new vertex located half the distance from the projected vertex to the centroid. The search is terminated when several consecutive projections fail to improve the solution.

This algorithm was criticized by Guin (1968) for three reasons

1. In Box's algorithm an upper and lower bound is specified for each variable. If a vertex violates the bounds on any variable, this violation is resolved by placing the vertex just inside the violated bounds. This may result in an undesirable sameness in the vertices which will slow the convergence of the algorithm. A remedy is to simply move infeasible vertices further into the feasible region.

2. An infinite loop exists when there is no improved solution along the line from the worst point through the centroid. Guin suggests counting the number of trial solutions in any one direction and changing directions when some limit is encountered.

3. An infinite loop exists if the centroid becomes infeasible. Therefore, testing the feasibility of the centroid is recommended. If the centroid is not feasible a new complex can be formed around the best vertex.

Guin's recommendations have very little effect on moderately sized well behaved problems. However, incorporating these modifications in the algorithm makes it possible to use the minimum number of vertices $(n+1)$ and may improve performance when the feasible region is not convex.

In a multiobjective setting the function to be optimized is the decision maker's unspecified value function. However, since complex search uses the function to be optimized only to identify the best and worst points, the decision maker can be queried for this information

directly.   The following procedure developed by Ringuest and
Gulledge (1985) describes a complex search as it might be applied to a
multiobjective problem

1.   Construct a complex of $p = 2n$ points consisting of a feasible point
     and $p-1$ additional points generated from random numbers and
     upper and lower bounds for each of the variables

     $$x_{ij} = L_i + r_{ij}(U_i - L_i); \; i = 1, 2, ..., n; \; j = 2, 3, ..., p$$
     where

          r_{ij} are pseudo-random numbers between 0 and 1

          L_i is the lower bound on variable i

          U_i is the upper bound on variable i.

2.   The selected points must satisfy both the bounds and constraints of
     the problem.   If at any time the bounds are violated, move the
     point a small distance δ inside the violated bound. If a constraint
     is violated, move the point one half of the distance to the
     centroid of the remaining points as follows

     $$x_{ij}(\text{new}) = [x_{ij}(\text{old}) + x_{ic}]/2; \; i = 1,2,...,n$$
     where

          $x_{ij}(\text{new})$ are the coordinates of the new point

          $x_{ij}(\text{old})$ are the coordinates of the infeasible point

          x_{ic} are the coordinates of the centroid of the remaining
               points, defined by

     $$x_{ic} = \frac{1}{p-1} [\sum_{j=1}^{p} x_{ij} - x_{ij}(\text{old})]; \; i = 1,2, ..., n.$$

     [Note: The sum is from $j = 1, 2, ..., p$ but does not include the
     infeasible point.]

Repeat this process until all of the constraints are satisfied.

3. Evaluate the objective functions at each point. Ask the decision
    maker to identify the point which yields the least preferred set of
    objective function values. Replace this point with a point located
    α times as far from the centroid of the remaining points as the
    distance from the rejected point to the centroid. The new point
    is placed along the line joining the rejected point and the
    centroid. Mathematically

    $$x_{ij}(new) = \alpha[x_{ic}-x_{ij}(old)]+x_{ic}; \ i=1, 2, ..., n.$$

    Box recommends using $\alpha = 1.3$.

4. If the new point x_j, repeats as the least preferred solution, move it
    one half the distance back toward the centroid of the remaining
    points.

5. Check the new point x_j against the bounds and constraints. If
    violations occur adjust the point as in step 2.

6. Terminate the algorithm when the decision maker discerns no
    improvement among the objective function values for
    consecutive iterations. An iteration is defined as the calculations
    required to select a new point which satisfies the constraints and
    does not repeat the least preferred solution.

    The algorithm as presented is suitable for relatively small (n
small) well behaved problems. Guin's recommendations can be
added to these steps to improve the performance of the algorithm
when the solution space is not convex and to allow for the use of the
smallest possible complex ($p=n+1$). When n is large so that there
are too many solution points for the decision maker to evaluate
unaided, the procedures described in the previous chapter (filtering,
clustering, matching and grouping or screening) may be used.

6.5   CHOOSING AN INTERACTIVE METHOD

A variety of approaches have been developed for multiple objective math programming problems. Yet, no one methodology or set of methodologies can be considered the "best" for all types of decision makers in all types of situations. Further, as pointed out by Gershon and Duckstein (1982) and Reeves and Franz (1985), it is not practical or desirable to impose a single multiobjective decision making technique as best for a given problem/decision maker/analyst. One characteristic of these algorithms which will likely influence their applicability is the burden placed on the decision maker at each step of the algorithm. Two algorithms have been discussed which are designed so that the decision maker is presented with relatively few alternatives at each stage of the algorithm and the decision maker is only asked to identify the least preferred solution.

A second factor which influences the applicability of multiobjective algorithms is computational efficiency. This will be discussed in detail in the next two chapters.

6.6   REFERENCES

Benayoun, R., J. de Montgolfier, J. Tergny and O. Laritchev, "Linear Programming with Multiple Objective Functions: Step Method (STEM)," *Mathematical Programming*, 1 (1972), 366-375.

Box, M. J., "A New Method of Constrained Optimization and a Comparison with Other Methods," *Computer Journal*, 8 (1965), 42-52.

Cohon, J. L., *Multiobjective Programming and Planning*, Academic Press, New York, 1978.

Cohon, J. L., R. L. Church and D. P. Sheer, "Generating Multiobjective Tradeoffs: An Algorithm for Bicriterion Problems," *Water Resources Research*, 15 (1979), 1001-1010.

Frank, M. and P. Wolfe, "An Algorithm for Quadratic Programming," *Naval Research Logistics Quarterly*, 3 (1956), 95-110.

Geoffrion, A. M., J. S. Dyer and A. Feinberg, "An Interactive Approach for Multicriterion Optimization, with an Application to the Operation of an Academic Department," *Management Science*, 19 (1972), 357-368.

Gershon, M. and L. Duckstein, "An Algorithm for Choosing a Multiobjective Technique," Fifth International Conference on Multiple Criteria Decision Making, Mons, Belgium, 1982.

Goicoechea, A., D. R. Hansen and L. Duckstein, *Multiobjective Decision Analysis with Engineering and Business Applications*, John Wiley and Sons, New York, 1982.

Guin, J. A., "Modification of the Complex Method of Constrained Optimization," *Computer Journal*, 10 (1968), 416-417.

Haimes, Y. Y. and W. A. Hall, "Multiobjectives in Water Resources Systems Analysis: The Surrogate Worth Trade-Off Method," *Water Resources Research*, 10 (1974), 615-623.

Hemming, T., "A New Method for Interactive Multiobjective Optimization: A Boundary Point Ranking Method," *Lecture Notes in Economics and Mathematical Systems*, No. 130, Berlin: Springer-Verlag, (1976), 333-339.

Korhonen, P. and J. Laakso, "Solving Generalized Goal Programming Problems Using a Visual Interactive Approach," *European Journal of Operational Research*, 26 (1986), 355-363.

Nelder, J. A. and R. Mead, "A Simplex Method for Function Minimization," *Computer Journal*, 7 (1965), 308-313.

Ringuest, J. L. and T. R. Gulledge, Jr. "Interactive Multiobjective Complex Search," *European Journal of Operational Research*, 19 (1985), 362-371.

Reeves, G. R. and L. S. Franz, "A Simplified Interactive Multiple Objective Linear Programming Procedure," *Computers and Operations Research*, 12 (1985), 589-601.

Spendley, W., G. R. Hext and F. R. Himsworth, "Sequential Applications of Simplex Designs in Optimization and Evolutionary Operation," *Technometrics*, 4 (1962), 441-461.

Steuer, R. E., "An Interactive Multiple Objective Linear Programming Procedure," *TIMS Studies in the Management Sciences*, 10 (1977), 225-239.

Steuer, R. E., *Multiple Criteria Optimization: Theory, Computation, and Applications*, John Wiley and Sons, New York, 1986.

Steuer, R. E. and E. F. Wood, "A Multiple Objective Markov Reservoir Release Policy Model," College of Business Administration, University of Georgia, Athens, Georgia, 1986.

Zionts, S. and J. Wallenius, "An Interactive Programming Method for Solving the Multiple Criteria Problem," *Management Science*, 22 (1976), 652-663.

Zionts S. and J. Wallenius, "An Interactive Multiple Objective Programming Method for a Class of Underlying Nonlinear Utility Functions," *Management Science*, 29 (1983), 519-529.

7

COMPUTATIONAL EFFICIENCY AND PROBLEMS WITH SPECIAL STRUCTURE

One common characteristic of many methods for solving multiobjective linear programming problems (particularly the methods for generating the nondominated set and the interactive methods) is that they require the solution of a sequence of single objective linear programming problems. In large scale problems (i.e. problems with many objectives, constraints or variables) this can result in a significant computational burden. This chapter will discuss ways in which the structure of the problem can be used to reduce this computational burden.

7.1 NETWORK FLOW PROBLEMS

Network flow problems may be defined as those problems that can be modeled as an interacting network of nodes and connecting links or arcs. Nodes often represent such physical entities as factories, warehouses, or customers. Arcs may be used to model the production at a factory, the inventory level in a warehouse, or the transporting of a product to a customer.

The network flow problem can be described as a linear programming problem of the form

Minimize: cx

Subject to: $Ax = b$

$x \geq 0$

where

A is an mxn matrix

b is an m dimension column vector

c is an n dimension row vector

x is an n dimension column vector

and the following properties also hold

1. each of the n arcs of the network corresponds with a column of **A**

2. each of the m nodes of the network corresponds with a row of **A**

3. if arc h is directed from node i_h to node j_h, then column h of **A** has a coefficient of -1 in row i_h, +1 in row j_h and zeros in the other m-2 rows.

When a network flow problem is formulated in this way, c is a vector of arc costs, **b** is a vector of net demands, and **x** is an n dimensional vector of arc flow variables. The matrix **A** has a very special structure and is referred to as a node-arc incidence matrix. Further, since **A** is unimodular [see Heller and Tompkins (1956) for the proof], every basic feasible solution to the network flow problem is integer if **b** is an integer vector.

Numerous algorithms have been developed for solving single objective network flow problems. These include network specializations of more general purpose algorithms [e.g. Barr et al. (1974) and Glover et al. (1972, 1974)] and new algorithms designed for special types of network flow problems [e.g. Dial et al. (1979) and Glover et al. (1979)]. Enormous computational gains have been obtained by taking advantage of the special properties of network flow problems. Many of the specialized network algorithms have exhibited from one to three orders of magnitude improvement in solution time over the more general purpose linear programming algorithms, depending on the degree of complexity of the network problem being solved. These computational gains should also be obtainable in multiple-objective network flow problems.

7.2   MULTIPLE-OBJECTIVE NETWORK FLOW PROBLEMS

The multiobjective network flow problem is defined as

Minimize $\mathbf{Cx}$

Subject to: $\mathbf{Ax = b}$

$$\mathbf{x} \geq \mathbf{0}$$

where

C is a kxn matrix of arc costs

A is an mxn node-arc incidence matrix

b is an m dimensional column vector of net demands

x is an n dimensional vector of arc flow variables.

The methods for solving this problem are similar in type to those presented for the more general multiobjective linear programming problem.   That is, an approximation to the decision maker's multiattribute value function may be specified and optimized, or all nondominated solutions may be computed and then screened by the decision maker, or an interactive algorithm may be applied. Examples of each of these approaches will be presented in the next few sections.

7.3   A NETWORK SPECIALIZATION OF A
MULTIOBJECTIVE SIMPLEX ALGORITHM

The three approaches to multiple-objective network flow problems all require an efficient multiobjective simplex algorithm as part of their solution process.   Klingman and Mote (1982) have presented a network specialization of the general multiobjective simplex algorithm of Yu and Zeleny (1975).   The Yu and Zeleny algorithm can be used to generate all nondominated extreme point solutions for a general multiobjective linear program. The algorithm proceeds from an initial nondominated solution by solving a nondominance subproblem for each adjacent basis. The initial nondominated solution can be obtained by optimizing any one of the objective functions. If any alternate optimal solutions exist they will include a nondominated solution or if the solution is unique it will be nondominated. The nondominance subproblem is of the form

Minimize: $\displaystyle\sum_{h=1}^{k} d_h$

Subject to: $Ax = b$

$$c_h x + d_h = c_h x^b, h = 1, 2, ..., k$$

$$d \geq 0$$

where

x^b is the adjacent basis being tested.

If $d = 0$ at the maximum, then x^b is efficient. The modification by Klingman and Mote is designed to take advantage of the structure of the multiobjective network flow problem. Since the initial nondominated solution is obtained by solving a single objective network flow problem and the nondominance subproblem contains the network flow constraints, the simple basis structure of these problems can be used to advantage. In particular, the basis tree representation and updating techniques that have been developed for single objective network flow problems can be used to reduce the computational effort involved in implementing the multiobjective simplex method.

Once the set of nondominated solutions has been identified, intra-set point generation can be used to generate points from the efficient edges and faces of the feasible region. Filtering, clustering, sectioning, matching and grouping or stochastic screening may then be used to aid the decision maker in choosing a most preferred solution.

7.4   COMPROMISE SOLUTIONS FOR THE MULTIOBJECTIVE NETWORK FLOW PROBLEM

Diaz (1978) presents an alternative procedure to generating all nondominated solutions. He suggests applying the compromise programming methods described previously in Chapter 4. Briefly, this approach is dependent upon specifying, a priori, a measure of the closeness of any compromise solution to the ideal solution. The ideal solution is defined in solution space as the point that would yield the optimal value for each objective function taken separately. The closeness measure is represented by what is termed a compromise function. The compromise function is used in place of the set of objective functions, reducing the problem to a single objective, and is optimized subject to the original constraints of the problem. Diaz suggests two possible forms for the compromise function. They are the linear compromise function

$$f(x) = \sum_{h=1}^{k} y_h$$

where

$$y_h = \frac{1}{c_h x_h^*} c_h x$$

x_h^* is the x which optimizes the hth objective function subject to $Ax=b$

and the quadratic compromise function

$$f(x) = \sum_{h=1}^{k} y_h^2$$

$$y_h = \frac{1}{c_h x_h^*} c_h x - 1$$

x_h^* is again the x which optimizes the hth objective function
   subject to $Ax = b$.

The compromise problem is then to

Maximize $f(x)$

Subject to: $Ax = b$

                   $x \geq 0$.

These two functional forms are used for a number of reasons.
Perhaps, foremost is that either objective function will yield a single
objective mathematical program that is easily optimized. Using the
linear compromise function results in a single objective network flow
problem, thus, the computational efficiencies associated with that
problem can be obtained. The quadratic compromise function leads
to a quadratic programming problem which also has well established
solution procedures although they are computationally more
burdensome. Diaz has demonstrated some desirable properties of
the solutions to these two problems. Since the linear compromise
problem fits the form of a standard network flow problem, the
solution (or one of the alternate optimal solutions) is nondominated
and occurs at an extreme point. The quadratic compromise problem
also yields a nondominated solution; in general, this solution will not
be an extreme point.

Compromise programming, as discussed in Chapter 3, is
applicable to the general multiple-objective linear programming
problem. The linear and quadratic compromise problems are both
completely objective. How well the solution to either of these
problems approximates the decision maker's value maximum solution
is dependent on how well the compromise function approximates the

decision maker's multiattribute value function. The decision maker provides no input to the solution process in either of these two approaches. In the next section, interactive solutions to the multiple objective network flow problem will be discussed.

7.5 INTERACTIVE METHODS FOR THE MULTIOBJECTIVE NETWORK FLOW PROBLEM

Klingman and Mote (1982) describe an interactive solution procedure for solving multiple-objective network flow problems. In their algorithm, the decision maker is asked to assign weights to each individual objective. A single weighted objective or "surrogate criterion" network problem is then solved. New weights are elicited and the process repeated until the decision maker is satisfied with the solution. As long as the weights are strictly positive, the resulting solution will be nondominated. However, two very different sets of weights may yield the same solution, thus this procedure may be tedious.

Ringuest and Rinks (1987) describe two algorithms which take advantage of the computational advances made by Klingman and Mote. These methods require less information from the decision maker than the surrogate criterion approach and will therefore be less burdensome. These algorithms will also accommodate two different decision making styles.

7.5.1 Algorithm One

To initialize an interactive algorithm for the multiobjective network flow problem it is necessary to determine a nondominated solution. It is desirable for this initial nondominated solution to be as close as possible to the value maximum solution. A number of researchers, including Edwards (1977) and Einhorn and McCoach (1977), advocate the use of an additive multiattribute value function incorporating linear single objective conditional value functions as an approximation to the "true" value function. Thus, an additive linear function, such as the linear compromise function, may provide a good initial solution. The set of nondominated solutions adjacent to the linear compromise solution can be easily obtained by solving the nondominance subproblem described earlier. The decision maker can then be called on to compare these solutions and a search analogous to sectioning can be initiated. These steps can be combined to form the algorithm that is described below.

1. Solve the k single objective problems

   Minimize $\qquad c_h x$, $h = 1, 2, ..., k$

   Subject to: $\qquad Ax = b$

   $\qquad\qquad\quad x \geq 0$

   and compute the k nondominated solutions $Z_h = [c_1 x_h^*, ..., c_k x_h^*]$ where x_h^* is the optimal extreme point for the hth objective.

2. Determine the optimal linear compromise solution and form the set of all nondominated solutions from steps 1 and 2.

3. Present the decision maker with the set of nondominated solutions and elicit the most preferred solution. If this solution was previously identified as the most preferred solution, go to step 7a.

4. Determine if the decision maker is satisfied with the solution identified in step 3.

   a. If the decision maker is satisfied, STOP.

b.   If the decision maker is not satisfied, continue with step 5.

5.   Determine all nondominated solutions that correspond with extreme point solutions that are adjacent to the extreme point identified in step 3, 7b, or 7c.

6.   Form the set of nondominated solutions consisting of all solutions from steps 3 and 5 and return to step 3.

7a. If there are no other nondominated solutions for the current iteration and there are no prior iteration(s), then have the decision maker identify the most preferred solution from the set of all nondominated solutions and STOP.

7b. If there are no other nondominated solutions for the current iteration and there are prior iteration(s), then retrieve the set of nondominated solutions from the previous iteration. Identify the most preferred solution that has not been examined in step 5 and return to step 5.

7c. If there are other nondominated solutions for the current iteration, then elicit the next most preferred solution from the decision maker and return to step 5.

Steps 1 and 2 in this algorithm determine the initial set of nondominated solutions. The solutions found in step 1 are included so that the algorithm begins with a set of solutions that spans the solution space. This will be important in the event that the linear compromise solution provides a poor approximation to the underlying value function and thus a poor starting point. Steps 3 through 6 incorporate the decision makers responses into the search procedure in a manner analogous to sectioning. Step 7 allows the algorithm to backtrack and provides the decision maker with the opportunity to reexamine some of the previous solutions.

7.5.2 Algorithm Two

The second algorithm presented by Ringuest and Rinks is a network modification of the SIMOLP procedure of Reeves and Franz (1985) which was presented in the previous chapter. Recall that the Reeves and Franz procedure optimizes a single weighted linear combination of the objectives (subject to the original constraints of the problem) at each stage of the algorithm. Thus, in this application the algorithm solves a single objective network flow problem at each stage and can take advantage of the problem structure to reduce the computational burden. The modified SIMOLP algorithm might proceed as follows (additional technical details are included in the earlier presentation):

1. Solve the k single objective problems and compute the k nondominated solutions.

2. Ask the decision maker to identify the most preferred nondominated solution.

3. Determine if the decision maker is satisfied with the solution identified in step 2.

    a. If the decision maker is satisfied, STOP.

    b. If the decision maker is not satisfied, continue with step 4.

4. Identify a function (hyperplane) which passes through the k current nondominated solutions.

5. Determine the vector x' which optimizes the function found in step 4 subject to the original constraints of the problem.

6. Present the decision maker with the Cx' solution. If Cx' is preferred to at least one solution in the current set of nondominated solutions and x' is new, proceed with step 7a. Otherwise, execute step 7b.

7a.   Add $\mathbf{C\alpha}'$ to the set of nondominated solution, delete the least preferred solution from this set, and return to step 4.

7b.   Have the decision maker identify the most preferred solution from the set of nondominated solution, STOP.

Step 1 in this procedure is identical to the previous algorithm; this step and step 2 initialize the algorithm with a set of nondominated solutions that spans the solution space. Steps 4 and 5 identify a linear combination of the problems objectives that is based on preference information elicited from the decision maker. The optimization of this linear combination will yield a nondominated solution. Steps 4 and 5 along with steps 6 and 7 are repeated iteratively until a most preferred solution to the problem is identified. This algorithm is also capable of backtracking since it may return to solutions that were removed in earlier steps.

7.5.3 A Comparison of Algorithms One and Two

Two interactive algorithms for the multiple-objective network flow problem have been described. The first algorithm presented begins in constructing a linear compromise solution. From this point, a search is conducted among all nondominated solutions corresponding with extreme points that are adjacent to the most preferred extreme point currently available. This search is continued until a satisfactory solution is obtained. Thus, the algorithm proceeds from one efficient extreme point to the next along the edges of the feasible decision variable space. Algorithm two proceeds by optimizing a function which passes through the k currently most preferred solutions. This procedure is repeated iteratively until a nondominated solution is repeated or a less preferred solution results. Thus, the algorithm is allowed to move back and forth across the objective function space.

The two algorithms also use different stopping criteria. Algorithm one requires that the decision maker express satisfaction with a nondominated solution while algorithm two terminates when a nondominated solution is repeated or a less preferred solution occurs. Either algorithm could be modified to include other termination schemes as neither is bound to their present criteria.

In both algorithms the number of iterations required to reach a satisfactory solution is dependent on the closeness of the solutions used to initialize the algorithm to the most preferred solution. The specific solutions used are chosen so that they will span the objective function space. It is hoped that at least one of the initial solutions will be in the proximity of the most preferred solution. Should this not be the case, the quality of the final solution identified will likely be influenced by the persistence of the decision maker. Thus, it will be important to match the manager's decision making style with the appropriate algorithm.

Algorithm one implies a more pro-active style since the decision maker is asked to identify the best solution at each step and termination does not occur until the decision maker is satisfied. Conversely, algorithm two deletes the worst alternative at each step and terminates when a solution is repeated or when the new solution generated is less preferred than all solutions in the current set. This algorithm would better match a more passive decision making style.

Perhaps the greatest disparity in these two interactive algorithms arises with respect to the information burden placed upon the decision maker. Algorithm one requires the decision maker to identify the most preferred solution from among the last most preferred solution and all nondominated solutions corresponding with immediately adjacent efficient extreme points. This is potentially as many as m*n-m-n+2 solutions, although the number will likely be smaller near the optimal solution. In algorithm two the decision maker will always be faced with exactly k+1 alternatives. Thus, unless a problem has a large number of objectives, the second algorithm imposes less of a burden on the decision maker.

Computationally both algortihms take advantage of the special structure of network flow problems. Algorithm one incorporates the multiobjective network flow simplex algorithm of Klingman and Mote (1982) within the solution procedure. Algorithm two requires the solution of a series of single objective network flow problems where any efficient algorithm can be used. Thus, the two algorithms should be efficient with respect to the computational burden required per iteration.

Computational efficiency is also a concern for the general multiobjective mathematical programming problem.   The next chapter will discuss computational issues in the more general context.

7.6   REFERENCES

Barr, R., F. Glover and D. Klingman, "An improved Version of the Out-of-Kilter Method and a Comparative Study of Computer Codes," *Mathematical Programming*, 7 (1974), 60-87.

Dial, R., F. Glover, D. Karney and D. Klingman, "A Computational Analysis of Alternative Algorithms and Labeling Techniques for Finding Shortest Path Trees," *Networks*, 9 (1979), 215-248.

Diaz, J. A., "Solving Multiobjective Transportation Problems," *Ekonomicko-Matematicky Obzor*, 14 (1978), 267-274.

Edwards, W., "How to Use Multiattribute Utility Measurement for Social Decision Making," *IEEE Transactions on Systems, Man, and Cybernetics*, 7 (1977), 326-340.

Einhorn, H. J. and W. McCoach, "A Simple Multiattribute Utility Procedure for Evaluation," *Behavioral Science*, 22 (1977), 270-282.

Glover, F., D. Klingman and A. Napier, "An Efficient Dual Approach to Network Problems," *OPSEARCH*, 9 (1972), 1-19.

Glover, F., D. Karney and D. Klingman, "Implementation and Computational Comparisons of Primal, Dual, and Primal-Dual Computer Codes for Minimum Cost Network Flow Problems," *Networks*, 4 (1974), 191-212.

Glover, F., D. Klingman, J. Mote and D. Whitman, "Comprehensive Computer Evaluation and Enhancement of Maximum Flow Algorithms," Research Report CCS 360, Center for Cybernetic Studies, The University of Texas at Austin, 1979.

Heller, I. and C. Tompkins, "An Extension of a Theorem of Dantzig's," in A. W. Tucker (ed.), *Linear Inequalities and Related Systems*, Princeton University Press, Princeton, New Jersey, 1956.

Klingman, D. and J. Mote, "Solution Approaches for Network Flow Problems with Multiple Criteria," *Advances in Management Studies*, 1 (1982), 1-30.

Reeves, G. R. and L. S. Franz,"A Simplified Interactive Multiple Objective Linear Programming Procedure," *Computers and Operations Research*, 12 (1985), 589-601.

Ringuest, J. L. and D. B. Rinks, "Interactive Solutions for the Linear Multiobjective Transportation Problem," *European Journal of Operational Research*, 32 (1987), 96-106.

Yu, P. L. and M. Zeleny, "The Set of All Nondominated Solutions in Linear Cases and a Multicriteria Simplex Method," *Journal of Mathematical Analysis and Applications*, 49 (1975), 430-468.

8

COMPUTATIONAL EFFICIENCY AND LINEAR
PROBLEMS OF GENERAL STRUCTURE

The previous chapter presented an initial discussion of computational issues related to solving large scale multiobjective linear programming problems. The discussion was restricted to problems which can be structured as multiobjective network flow problems. This chapter will broaden the discussion to linear programming problems with general structure. The discussion will describe a study of the computational requirements for obtaining a representation of the ideal solution for large scale multiobjective linear programs.

8.1 COMPUTATIONAL EFFICIENCY AND THE IDEAL SOLUTION

In Section 4.1 we defined the ideal solution and described its role in the human decision making process. Haksever and Ringuest (1989) have conducted a study which focuses on computational difficulties involved in obtaining a representation of the ideal solution for large scale problems. In their study the number of iterations and CPU time required to obtain a representation of the ideal solution are used as the basis for comparing four approaches to obtaining this representation. The emphasis is placed on the ideal solution: 1) because of the importance of this solution as described in Section 4.1, and 2) so that the comparisons in the study would be independent of any specific multiobjective linear programming algorithm.

As an initial benchmark for comparison a naive approach was used. A naive way to obtain a representation of the ideal solution is to solve each of the k single objective linear programs independently, i. e.

Maximize	$Z_h(\mathbf{x})$
Subject to:	$\mathbf{Ax} = \mathbf{b}$
	$\mathbf{x} \geq 0$

for all h, starting each problem from the origin. With large problems (particularly problems with a large number of constraints) it may take many iterations just to obtain a feasible solution. Using this naive approach, these steps must be retraced for each and every problem. The second approach used was designed to attack this problem.

In solving these k linear programs where each linear program has the identical feasible region, it seems desirable to remain in the feasible region once a feasible solution has been found. Remaining in the feasible region will eliminate repetition of the steps needed to reach the feasible region for each of the 2nd through k^{th} problems. This idea was implemented by using the optimal solution to the hth problem as the initial solution to the $h+1^{th}$ problem, for h = 1, ..., k-1. Armstrong, Charnes, and Haksever (1987, 1988) have applied a similar method to non-convex ratio goals problems.

Using this approach a large number of steps may still be repeated. Suppose the hth and $h+1^{th}$ objectives happen to be optimized at distant corners of the feasible region. Then using the optimal solution to objective h as the initial solution for objective $h+1$ will cause the simplex procedure to traverse many corners. Now suppose objective $h+2$ is optimized near objective h. Using the optimal solution for objective $h+1$ as the initial solution for objective $h+2$ will again cause the simplex method to (re)traverse many corners. The third and fourth approaches implemented were designed to reduce the likelihood of this occurring.

The problem just described suggests that there is an optimal order in which the k single objective linear programs should be solved. Specifically, they should be solved in an order such that the minimum total number of corner points are traversed in moving from the first to the kth optimal solution. Additionally, in order to minimize computer storage requirements each optimal corner should serve as the initial corner for one and only one linear program. In this way there is no increase in the number of bases that are stored. These requirements describe the search for a Hamiltonian path. Thus, the authors consider heuristics based on this problem and the related travelling salesman problem.

To model a travelling salesman problem the distance between each pair of possible destinations must be known. In this application the relevant distance is the number of extreme points between each of the k optimal solutions. This distance, however, is not known until all k problems are solved. Therefore, a proxy measure is needed. One of the primary factors which determines the closeness of two optimal solutions (along with the number, inclination and direction of the constraints) is the inclination of the two corresponding objective functions. The inclination can be conveniently described by specifying the normal to the objective function hyperplane. Since each objective is linear the corresponding normals are simply the objective function coefficients of each objective. Once the normals to the k objectives are specified the unique angle between any two normals, θ, can be determined. This angle can easily be characterized by the cosine, $\cos\theta$, which is defined as

$$\cos\theta = <u, v> / \|u\| \|v\|$$

where u and v are nonzero vectors, $<u, v>$ is the inner product of u and v, and $\|u\|$ is the norm of u. The $\cos\theta$ is then used as a proxy for the number of extreme points between each pair of objectives. Steuer (1986) uses the angle, θ, as a measure of the correlation between two objectives.

The travelling salesman problem is a difficult, NP hard, combinatorial optimization problem. [*The Travelling Salesman Problem*; Lawler, Lenstra, Rinnooy Kan, and Shmoys; editors (1985) provides a historical as well as up to date perspective on the state of combinatorial optimization.] Due to the difficulty inherent in obtaining optimal solutions for this type of problem many heuristics have been developed which find good approximate solutions. Golden, Bodin, Doyle and Stewart (1980), and Golden and Stewart (1985) have conducted studies of the computational efficiency and accuracy of some of these heuristics.

For this application, the heuristic chosen had to possess two particular properties. First, it needed to be computationally efficient. The ultimate objective was to increase computational efficiency in obtaining a representation of the ideal solution. Thus, the computational advantage gained by ordering the objectives in the multiobjective linear program could not be lost to a computationally inefficient travelling salesman heuristic. Secondly, the heuristic needed to be reasonably accurate. Since a proxy measure is used for the true distance (number of extreme points) between optimal solutions it did not make sense to use a heuristic that spent a lot of computations on making small improvements in the accuracy of the solution.

Golden et al. showed that many of the tour construction procedures (procedures which generate an approximately optimal solution from the distance matrix) will find a solution within 5% to 7% of optimal. This is also the most computationally efficient class of solution procedures that they tested.   Of these Haksever and Ringuest selected two methods developed by Rosenkrantz, Stearns and Lewis (1974); namely, arbitrary insertion and nearest neighbor. These two procedures were selected because they require the fewest computations (on the order of k^2). Arbitrary insertion was found to be the most accurate (Golden et al.), and nearest neighbor is the most intuitive.

The arbitrary insertion algorithm that was implemented can be described by the following steps

1. Start with a subgraph consisting of node i (objective i) only.
2. Find node h such that $c_{ih} = \cos(\theta_{ih})$ [θ_{ih} is the angle between the normals to objectives i and h] is maximum and form subtour i-h-i.
3. Given a subtour, arbitrarily select node h not in the subtour.
4. Find the arc (i,j) in the subtour which maximizes $c_{ih} + c_{hj} - c_{ij}$. Insert h between i and j.
5. Go to step 3, unless we have a Hamiltonian cycle.
6. Eliminate the longest arc.

Notice that the largest cosine is chosen since the largest cosine implies the smallest angle and that the longest arc is eliminated since once the last node (objective) is reached it is not necessary to go back to the first node. Golden et al. derived the worst case behavior of this procedure

arbitrary insertion tour length/optimal tour length $\leq 2\ln(k) + 0.16$.

Their computational results found this heuristic to be accurate within 5% of optimal.

The nearest neighbor algorithm was implemented as follows

1. Start with any node as the beginning of a path.

2. Find the node closest to the last node added to the path [i.e. find the objective with the normal that makes the smallest angle (largest cosine) with the most recently added objective's normal]. Add this node to the path.

3. Repeat step 2. until all nodes are contained in the path.

Worst case behavior for the nearest neighbor algorithm is

nearest neighbor tour length/optimal tour length $\leq 1/2 \, [\lg(k)] + 1/2$

(Golden et al.). The computational results of Golden et al. found this algorithm to be accurate within 17% of optimal.

8.2   TEST PROBLEMS

The four approaches described above were implemented and tested on a set of multiobjective linear programming (MOLP) problems. A program entitled, Single or Multiobjective Linear Programming Problem Generator (SOMOLPG), was developed and coded in standard FORTRAN. This program is largely based on the method proposed by Charnes, Raike, Stutz, and Walters (1974) for generating linear programming (LP) problems with known solutions. SOMOLPG is capable of randomly generating linear mathematical programming problems with specified characteristics. Problem parameters such as: the number of objectives, the number of variables and constraints, and the problem density are supplied by the user. The data structure of the problems is compatible with the SMU-LP computer code [Ali and Kennington (1980)].

Three sets of MOLP problems with approximate densities of 0.25, 0.15, and 0.05 were generated for testing purposes. In each set, five fixed problem sizes (ranging from 200 variables and 40 constraints to 1000 variables and 200 constraints) were used in generating MOLP problems with 2, 4, 6, 8, and 10 objectives. Each of these problems was generated independently, resulting in a total of 75 distinct random problems that were used for testing each of the four approaches. One problem characteristic that was constant for all problems was the ratio (0.2) of the number of constraints to the number of variables. Representations of the ideal solutions were obtained for each of the 75 test problems using the four different approaches.

8.3   COMPUTER CODES

Central to the computational testing reported by Haksever and Ringuest was SMU-LP a large scale LP code in standard FORTRAN [Ali and Kennington (1980)]. This code was first modified to implement the naive approach for obtaining a representation of the ideal solution. The modified code solved each of the k single objective LP problems starting from the origin. This modified code was named MOLP-N for MOLP Naive approach.

A second code developed from SMU-LP was MOLP-E for MOLP Efficient approach. This code was designed to solve the k single objective LP problems in the order in which they are generated without returning to the origin to begin each new problem. After the first problem ($h = 1$), the optimal basis and optimal solution of the most recently solved problem are kept while the coefficients of the hth objective function are replaced by those of objective $h + 1$ in the problem data. Once this operation is completed, the new problem starts at an advanced stage of the simplex procedure. Specifically, problem h+1 starts from the optimal basic solution to problem h. This point will always be a basic feasible solution since both problems have exactly the same feasible region.

The motivation for the third and fourth approaches described in Section 8.1 is the conjecture that the order in which the k single objective LP problems are solved can greatly influence the number of simplex iterations required. One way to test this hypothesis is to compare the number of simplex iterations over all possible orderings for the k single objective LPs. Since there are k! different orderings for each test problem, Haksever and Ringuest compared a subset of the test problems. Table 8.1 shows the minimum, maximum and average number of simplex iterations for 15 test problems with 4 objectives (problem characteristics are described in detail in a later section). These results clearly demonstrate that the order in which the k single objective LPs are solved effects the total number of simplex iterations.

To implement the third approach described above which determines the order in which the k objectives should be optimized using the arbitrary insertion algorithm, a subroutine was developed and added to MOLP-E. This version of MOLP-E was called MOLP-AI. The subroutine first computes the cosines of the angles between the normals of each pair of objective functions. Then, using the cosine as a proxy for the distance between nondominated extreme points, the subroutine uses the arbitrary insertion algorithm to determine an approximately optimal order for solving the k single objective LPs. The objective functions are then optimized in the prescribed order using MOLP-E.

The fourth approach involved determining the order in which the objective functions should be optimized using the nearest

Table 8.1 Variability in the Number of Simplex Iterations Required to Obtain a Representation of the Ideal Solution Due to the Order in Which the Problems are Solved - Problems with 4 Objectives

| | Number of Simplex Iterations | | |
Problem Number	Minimum	Maximum	Average
01-4	788	1033	903.50
11-4	646	938	769.50
21-4	168	227	203.00
02-4	2509	3692	3155.00
12-4	2671	3509	3008.75
22-4	1055	1661	1341.75
03-4	3912	5504	4818.75
13-4	3508	4637	4088.25
23-4	3789	5062	4515.50
04-4	7179	10716	8884.75
14-4	6290	8810	7779.75
24-4	7585	9431	8530.00
05-4	10423	13426	11974.50
15-4	7752	9827	8869.00
25-4	9113	12023	10485.25

neighbor algorithm. This was also implemented as a subroutine added to MOLP-E. The modified code was termed MOLP-NN. The subroutine begins with the first objective function (h = 1). It then computes the cosine of the angle between the normal of the last objective function added to the sequence and the normal of each objective function not yet in the sequence. The objective function which forms the smallest angle is the next one added to the sequence. When all k objective functions are included in the sequence MOLP-E is used to solve the k LPs in the order specified.

These four codes are identical in every respect except as mentioned above. Computational testing consisted of generating one problem at a time and then solving it by each of the four codes in succession. The solution statistics in Tables 8.1-8.6 are only for obtaining a representation of the ideal solution and the time statistics in Tables 8.2-8.6 do not include input/output times. All of the problems reported by Haksever and Ringuest were generated and solved on the VAXCluster system of Boston College using double precision arithmetic, a VAX FORTRAN compiler Version 4 and a VAX8700 computer.

8.4   RESULTS

Solution statistics for each of the 75 problems solved and each of the 4 computer codes used are presented in Tables 8.2 through 8.6. Each table includes statistics for 15 problems with a specific problem size ranging from problems with 200 variables and 40 constraints in Table 8.2 to problems with 1000 variables and 200 constraints in Table 8.6. Entries within each table are presented in five groups with three rows in a group. The first three rows in each table are for problems with 2 objective functions, the next three rows are problems with 4 objective functions, and so on up to the last three rows which are problems with 10 objective functions. Within each group of three problems is one problem with approximately 25% density, one with approximately 15% density and one with approximately 5% density. The number of iterations and the time in CPU seconds required to obtain a representation of the ideal solution to each problem is presented for each of the computer codes used. These entries are followed by the mean number of simplex iterations and mean CPU time required to obtain a representation of the ideal solution for all problems in the table solved by computer codes; MOLP-E, MOLP-AI and MOLP-NN.

ANOVAs with computer code as treatment factor, number of objectives and density as blocking factors, and number of simplex iterations and time as dependent variables were run. Both blocking factors were significant in all cases. The last entries in Tables 8.2 through 8.6 present the results of Scheffe's multiple comparison test applied to these means. Means with the same letter are not significantly different ($\alpha = .05$).

These tables show that consistent and substantial savings in both the number of iterations and the time required to obtain a representation of the ideal solution were achieved using the MOLP-E code. For problems with 2 objective functions both measures were reduced by approximately 20%. These reductions increased to approximately 60% for problems with 10 objective functions.

Tables 8.2 through 8.6 provide some evidence that additional savings in both the number of simplex iterations and time required can be achieved using a traveling salesman heuristic (MOLP-AI or MOLP-NN) to determine the order in which the k LP problems are

Table 8.2 Solution Statistics for Problems with 200 Variables and 40 Constraints

Problem Number	Dens.	Number of Iterations			
		MOLP -N	MOLP -E	MOLP -AI	MOLP -NN
01-2	.2524	662	446	446	446
11-2	.1554	769	505	505	505
21-2	.0550	289	233	233	233
01-4	.2612	1520	878	788	859
11-4	.1449	1165	668	702	668
21-4	.0551	367	182	208	174
01-6	.2466	2008	989	1068	999
11-6	.1578	2097	849	859	864
21-6	.0555	630	279	244	268
01-8	.2554	2385	916	863	912
11-8	.1510	2388	1108	1057	1197
21-8	.0579	791	269	253	258
01-10	.2491	3450	1929	1524	1441
11-10	.1535	2503	1048	881	875
21-10	.0557	1019	320	331	293
Mean			707.9	666.1	664.1
Scheffe Groups			A	A	A

Table 8.2 Cont. Solution Statistics for Problems with 200 Variables and 40 Constraints

Problem Number	Dens.	Time (CPU Seconds)			
		MOLP -N	MOLP -E	MOLP -AI	MOLP -NN
01-2	.2524	5.82	3.98	3.95	4.07
11-2	.1554	5.39	3.60	3.61	3.67
21-2	.0550	0.34	0.32	0.31	0.34
01-4	.2612	13.51	8.09	7.26	8.16
11-4	.1449	7.75	4.47	4.27	4.56
21-4	.0551	0.36	0.24	0.27	0.24
01-6	.2466	19.47	9.64	9.11	8.55
11-6	.1578	14.62	6.17	5.99	6.15
21-6	.0555	0.66	0.40	0.37	0.39
01-8	.2554	20.97	8.70	8.24	9.02
11-8	.1510	15.76	7.31	6.89	8.33
21-8	.0579	0.93	0.52	0.38	0.47
01-10	.2491	29.76	18.03	13.70	13.30
11-10	.1535	15.59	7.00	5.63	5.92
21-10	.0557	1.05	0.52	0.53	0.54
Mean			5.26	4.91	4.70
Scheffe Groups			A	A	A

Table 8.3 Solution Statistics for Problems with 400 Variables and 80 Constraints

		Number of Iterations			
Problem Number	Dens.	MOLP -N	MOLP -E	MOLP -AI	MOLP -NN
02-2	.2493	2073	1658	1658	1658
12-2	.1438	2227	1825	1825	1825
22-2	.0530	1049	891	891	891
02-4	.2421	4790	3182	3051	2509
12-4	.1565	4283	3041	2899	2712
22-4	.0482	1847	1359	1443	1236
02-6	.2438	6461	3723	3647	3376
12-6	.1546	6525	3710	3382	3502
22-6	.0505	4203	2684	2444	2325
02-8	.2432	9491	4369	4331	3847
12-8	.1442	8329	3900	3319	3342
22-8	.0480	5128	2666	2492	2474
02-10	.2529	10044	4410	4413	3932
12-10	.1466	10486	3965	3687	3772
22-10	.0503	5725	2651	2561	2323
Mean			2935.6	2802.9	2648.3
Scheffe			A	A	
Groups				B	B

**Table 8.3 Cont. Solution Statistics for Problems with 400 Variables
and 80 Constraints**

Problem Number	Dens.	Time (CPU Seconds)			
		MOLP -N	MOLP -E	MOLP -AI	MOLP -NN
02-2	.2493	66.86	56.58	59.83	57.89
12-2	.1438	55.99	48.60	49.03	51.08
22-2	.0530	10.98	10.36	9.83	10.14
02-4	.2421	158.79	111.25	106.25	88.87
12-4	.1565	107.04	80.57	75.45	72.46
22-4	.0482	14.52	11.62	12.16	10.75
02-6	.2438	206.86	125.06	124.58	118.66
12-6	.1546	167.57	98.05	91.60	95.27
22-6	.0505	44.11	30.64	26.85	25.82
02-8	.2432	279.41	144.33	142.52	133.76
12-8	.1442	197.89	94.79	80.41	81.21
22-8	.0480	50.98	26.37	24.47	25.80
02-10	.2529	320.79	153.37	159.86	141.37
12-10	.1466	247.49	100.71	94.10	95.45
22-10	.0503	59.78	26.44	25.60	23.65
Mean			74.60	72.17	68.81
Scheffe Groups			A	A	A

Table 8.4 Solution Statistics for Problems with 600 Variables and 120 Constraints

		Number of Iterations			
Problem Number	Dens.	MOLP -N	MOLP -E	MOLP -AI	MOLP -NN
03-2	.2458	4300	3284	3284	3284
13-2	.1437	4238	3657	3657	3657
23-2	.0521	2627	2155	2155	2155
03-4	.2468	8657	4275	4117	4275
13-4	.1519	7847	3508	3799	3685
23-4	.0508	6533	4541	3942	4541
03-6	.2429	12162	6713	6412	6448
13-6	.1513	11856	5124	4951	5050
23-6	.0482	8944	4649	4310	4252
03-8	.2534	18838	7372	6905	7035
13-8	.1495	16598	6193	5550	5548
23-8	.0512	14521	6430	6283	6079
03-10	.2508	21693	8700	8010	7739
13-10	.1488	19278	6493	5434	5455
23-10	.0497	14357	7133	5823	5896
Mean			5348.5	5006.6	4975.5
Scheffe Groups			A	A	A

Table 8.4 Cont. Solution Statistics for Problems with 600 Variables and 120 Constraints

Problem Number	Dens.	Time (CPU Seconds)			
		MOLP -N	MOLP -E	MOLP -AI	MOLP -NN
03-2	.2458	345.20	276.95	279.82	300.78
13-2	.1437	248.33	227.43	237.60	247.89
23-2	.0521	74.48	65.79	66.27	71.68
03-4	.2468	684.55	363.19	344.40	379.29
13-4	.1519	474.00	219.17	236.60	241.88
23-4	.0508	176.35	127.50	106.91	130.83
03-6	.2429	936.68	543.15	590.71	574.24
13-6	.1513	728.70	328.69	315.62	334.07
23-6	.0482	230.78	120.65	107.39	114.22
03-8	.2534	1527.52	648.50	594.06	615.03
13-8	.1495	1014.03	412.28	369.11	388.19
23-8	.0512	453.73	177.01	172.53	183.30
03-10	.2508	1682.51	747.06	664.95	661.00
13-10	.1488	1152.02	421.65	351.26	371.21
23-10	.0497	365.76	188.14	150.22	160.39
Mean			324.48	318.27	305.83
Scheffe Groups			A	A	A

Table 8.5 Solution Statistics for Problems with 800 Variables and 160 Constraints

Problem Number	Dens.	Number of Iterations			
		MOLP -N	MOLP -E	MOLP -AI	MOLP -NN
04-2	.2532	7505	6331	6331	6331
14-2	.1483	6796	5238	5238	5238
24-2	.0519	5644	4928	4928	4928
04-4	.2556	15348	7380	7798	7380
14-4	.1515	12781	7535	6290	7535
24-4	.0495	11638	8125	7585	8867
04-6	.2531	18258	6487	6629	7008
14-6	.1456	19942	8698	8139	7815
24-6	.0512	15903	6877	7479	6312
04-8	.2408	25292	10061	8858	9029
14-8	.1470	27185	10922	10519	9317
24-8	.0502	20442	9507	8802	8743
04-10	.2426	34226	7182	7434	6801
14-10	.1461	30401	12180	10845	10498
24-10	.0508	28843	9177	9657	9347
Mean			8041.9	7768.8	7676.6
Scheffe Groups			A	A	A

Table 8.5 Cont. Solution Statistics for Problems with 800 Variables and 160 Constraints

Problem Number	Dens.	MOLP -N	MOLP -E	MOLP -AI	MOLP -NN
			Time (CPU Seconds)		
04-2	.2532	1241.24	1090.69	1129.08	1124.82
14-2	.1483	877.47	723.99	719.80	744.41
24-2	.0519	327.15	292.93	307.08	302.51
04-4	.2556	2407.37	1263.60	1313.19	1323.86
14-4	.1515	1546.41	1037.65	844.38	1045.18
24-4	.0495	622.58	487.22	424.36	541.29
04-6	.2531	2865.57	1124.63	1135.17	1254.71
14-6	.1456	2480.14	1152.18	1058.08	1063.57
24-6	.0512	911.01	385.60	425.05	381.98
04-8	.2408	3763.90	1580.64	1418.68	1474.66
14-8	.1470	3287.10	1527.85	1391.07	1276.40
24-8	.0502	1122.41	543.89	500.74	541.13
04-10	.2426	5510.22	1232.29	1269.78	1191.83
14-10	.1461	3731.79	1609.62	1445.47	1469.08
24-10	.0508	1570.61	502.65	531.03	527.03
Mean			970.36	950.83	927.53
Scheffe Groups			A	A	A

Table 8.6 Solution Statistics for Problems with 1000 Variables and 200 Constraints

Problem Number	Dens.	Number of Iterations			
		MOLP -N	MOLP -E	MOLP -AI	MOLP -NN
05-2	.2447	10654	8561	8561	8561
15-2	.1457	10860	9224	9224	9224
25-2	.0499	9034	7581	7581	7581
05-4	.2505	20752	10651	10822	10981
15-4	.1492	20485	10008	10319	10040
25-4	.0509	17111	9113	9396	9113
05-6	.2406	32750	15446	13890	12507
15-6	.1460	28135	15661	13682	12575
25-6	.0514	25334	16110	13179	12359
05-8	.2530	45609	18180	18070	14562
15-8	.1516	39169	19783	15905	14864
25-8	.0506	32693	18127	16382	14923
05-10	.2552	51127	22082	19457	20643
15-10	.1513	51460	20059	18867	18816
25-10	.0490	41853	18736	17101	16171
Mean			14621.5	13495.7	12861.3
Scheffe			A		
Groups				B	B

Table 8.6 Cont. Solution Statistics for Problems with 1000 Variables and 200 Constraints

Problem Number	Dens.	Time (CPU Seconds)			
		MOLP -N	MOLP -E	MOLP -AI	MOLP -NN
05-2	.2447	2933.87	2511.62	2474.81	2527.35
15-2	.1457	2336.32	2082.27	2082.30	2206.29
25-2	.0499	944.28	870.36	838.69	861.34
05-4	.2505	5546.33	3139.62	3147.44	3307.92
15-4	.1492	4597.18	2437.53	2443.44	2452.62
25-4	.0509	2067.46	1007.22	1018.79	1056.31
05-6	.2406	8867.02	4664.62	4158.07	3832.17
15-6	.1460	6049.19	3647.33	3196.35	2933.42
25-6	.0514	2653.33	1839.15	1447.59	1377.70
05-8	.2530	12497.59	5396.00	5341.42	4477.18
15-8	.1516	8680.48	4831.69	3811.92	3583.91
25-8	.0506	3168.20	1864.70	1680.50	1593.75
05-10	.2552	13848.41	6964.98	5955.25	6623.52
15-10	.1513	10734.56	4637.95	4252.92	4464.31
25-10	.0490	3913.36	1850.85	1748.53	1599.26
Mean			3183.06	2906.53	2859.80
Scheffe Groups			A	A	A

solved. In every case, the mean number of simplex iterations and mean CPU time for MOLP-AI was less than MOLP-E and MOLP-NN was less than MOLP-AI.   However, these differences were statistically significant ($\alpha = .05$) only for the mean number of simplex iterations and problems with 400 variables or 1000 variables. Specifically, the mean number of simplex iterations for MOLP-NN was significantly less than the mean number of simplex iterations for MOLP-E in both cases and the mean number of simplex iterations for MOLP-AI was less than the mean number of simplex iterations for MOLP-E for problems with 1000 variables. In addition, Table 8.1 showed that there is a substantial difference between the minimum and maximum number of simplex iterations required depending on the order in which the K LP problems are solved. Comparing Tables 8.2 through 8.6 with Table 8.1 suggests that using a traveling salesman heuristic to order the objectives will at least smooth out this variability.

To further examine the computational efficiency of the MOLP-E, MOLP-AI, and MOLP-NN codes the number of iterations and CPU time required by each of these codes were expressed as a percent of that required by the naive approach, MOLP-N. Tables 8.7, 8.8 and 8.9 summarize the results of this analysis. Table 8.7 presents the mean and standard deviation over all problem sizes with approximate density of 25% segregated by number of objectives. The same information for problems with 15% density and 5% density are presented in Tables 8.8 and 8.9, respectively.

The small standard deviations in these tables show that there is little difference in computational efficiency (as measured in this study) that can be attributed to problem size. Tables 8.7, 8.8, and 8.9 show very similar results across the three levels of density. The magnitude of the results are similar across all three tables and the pattern of results is very consistent. For all densities and number of objectives the codes which incorporate a travelling salesman heuristic (MOLP-AI and MOLP-NN) perform on average as well or better than MOLP-E. For all densities and 6 or more objectives MOLP-AI and MOLP-NN perform on average 4% better than MOLP-E.

The results obtained by Haksever and Ringuest are so consistent, particularly with respect to problem size and density, that inference to other problems seems reasonable. With respect to the number of objectives, the range of values included in their study would seem to be comprehensive. Keeney and Raiffa (1976) present a number of case studies in multiobjective analysis. In each, the number of objectives falls within the range of this study. In addition, a strong pattern of results related to the number of objectives is established.

Two avenues would seem available for improving on the results obtained by Haksever and Ringuest. First, more exact solution procedures for the travelling salesman problem could be applied. The heuristics used were chosen primarily because they presented the least computational burden. Thus, any gains achieved by solving the LPs in a more appropriate order would not be lost in the computational burden of determining the order. A better (more accurate) heuristic or an exact solution procedure such as dynamic programming might improve the ordering but the ultimate results in terms of obtaining a representation of the ideal solution are not likely to be substantially improved. Golden et al. (1990) found the arbitrary insertion algorithm to be accurate within 5% of optimal and the nearest neighbor algorithm to be accurate within 17% of optimal yet the results for obtaining a representation of the ideal solution are not

Table 8.7 Measures of Computational Efficiency for Problems with
Approximate Density = .25

Number of Objectives		Number of Iterations as % of MOLP-N		
		MOLP-E	MOLP-AI	MOLP-NN
2	Mean	0.77	0.77	0.77
	Std. Dev.	0.065	0.065	0.065
4	Mean	0.54	0.53	0.52
	Std. Dev.	0.076	0.061	0.036
6	Mean	0.49	0.48	0.46
	Std. Dev.	0.085	0.088	0.076
8	Mean	0.41	0.39	0.37
	Std. Dev.	0.031	0.044	0.033
10	Mean	0.41	0.37	0.35
	Std. Dev.	0.13	0.090	0.089

**Table 8.7 Cont. Measures of Computational Efficiency for Problems
with Approximate Density = .25**

Number of Objectives		Solution Time (CPU) as % of MOLP-N		
		MOLP-E	MOLP-AI	MOLP-NN
2	Mean	0.81	0.83	0.84
	Std. Dev.	0.078	0.092	0.081
4	Mean	0.59	0.57	0.57
	Std. Dev.	0.070	0.063	0.026
6	Mean	0.52	0.51	0.50
	Std. Dev.	0.085	0.097	0.085
8	Mean	0.44	0.42	0.41
	Std. Dev.	0.043	0.054	0.045
10	Mean	0.45	0.40	0.40
	Std. Dev.	0.14	0.10	0.10

Table 8.8 Measures of Computational Efficiency for Problems with Approximate Density = .15

Number of Objectives		Number of Iterations as % of MOLP-N		
		MOLP-E	MOLP-AI	MOLP-NN
2	Mean	0.79	0.79	0.79
	Std. Dev.	0.082	0.082	0.082
4	Mean	0.56	0.55	0.55
	Std. Dev.	0.10	0.087	0.065
6	Mean	0.48	0.45	0.44
	Std. Dev.	0.077	0.051	0.058
8	Mean	0.44	0.39	0.39
	Std. Dev.	0.056	0.040	0.068
10	Mean	0.39	0.34	0.34
	Std. Dev.	0.030	0.036	0.036

Table 8.8 Cont. Measures of Computational Efficiency for Problems with Approximate Density = .15

Number of Objectives		Solution Time (CPU) as % of MOLP-N		
		MOLP-E	MOLP-AI	MOLP-NN
2	Mean	0.84	0.84	0.88
	Std. Dev.	0.098	0.11	0.12
4	Mean	0.60	0.57	0.60
	Std. Dev.	0.11	0.082	0.080
6	Mean	0.51	0.47	0.47
	Std. Dev.	0.083	0.065	0.060
8	Mean	0.48	0.41	0.42
	Std. Dev.	0.054	0.033	0.061
10	Mean	0.42	0.37	0.38
	Std. Dev.	0.030	0.036	0.037

Table 8.9 Measures of Computational Efficiency for Problems with
Approximate Density = .05

Number of		Number of Iterations as % of MOLP-N		
Objectives		MOLP-E	MOLP-AI	MOLP-NN
2	Mean	0.84	0.84	0.84
	Std. Dev.	0.024	0.024	0.024
4	Mean	0.63	0.63	0.63
	Std. Dev.	0.11	0.092	0.12
6	Mean	0.53	0.49	0.47
	Std. Dev.	0.10	0.070	0.058
8	Mean	0.47	0.43	0.42
	Std. Dev.	0.084	0.072	0.058
10	Mean	0.41	0.39	0.36
	Std. Dev.	0.087	0.051	0.055

**Table 8.9 Cont. Measures of Computational Efficiency for Problems
with Approximate Density = .05**

Number of Objectives		Solution Time (CPU) as % of MOLP-N		
		MOLP-E	MOLP-AI	MOLP-NN
2	Mean	0.92	0.91	0.94
	Std. Dev.	0.026	0.026	0.036
4	Mean	0.69	0.67	0.71
	Std. Dev.	0.12	0.13	0.13
6	Mean	0.59	0.53	0.52
	Std. Dev.	0.12	0.061	0.071
8	Mean	0.51	0.45	0.48
	Std. Dev.	0.077	0.059	0.046
10	Mean	0.45	0.43	0.42
	Std. Dev.	0.077	0.062	0.062

substantially different for these two heuristics. This suggests that other solution procedures for the travelling salesman problem will provide similar results in this application.

The second area for potential improvement is related to measuring the distance between the optimal solutions to two successive LP problems. The relevant distance measure is the number of corner points between two successive solutions. In the Haksever and Ringuest study, the angle between the norms of any two objective functions is used as a proxy for the distance between the optimal corner points for the two objective functions. This proxy is computationally very efficient but it only considers the inclination of the two objective functions. A better distance proxy would also consider the number, inclination and direction of the constraints. Constructing a proxy which considers the shape of the feasible region does not seem an easy task but may be worth further consideration.

Other methods may be available for improving MOLP algorithms which are related to the feasible region rather than the set of objective functions. Zionts and Wallenius (1983) have obtained some computational savings in integer problems by removing redundant constraints. Cheng (1981, 1985) has developed a set of theorems for identifying permanent basic and nonbasic variables which can be used to detect redundancy in linear programming. Haksever (1987) has tested the computational efficiency of these theorems as they are implemented to solve ratio goals problems using the Charnes-Cooper linearization method [Charnes and Cooper (1977)]. His study showed that this method yields little in computational efficiency. However, other approaches may merit investigation.

8.5   OTHER COMPUTATIONAL STUDIES

Haksever and Ringuest (1989) have also studied similar approaches for improving the computational efficiency of the SIMOLP procedure [Reeves and Franz (1985)]. The results they obtained are similar to those presented here. Very little other computational testing has been conducted. This is an area in need of additional study.

8.6   REFERENCES

Ali, A. I., and J. L. Kennington, "SMU-LP an In-core Primal Simplex Code for Solving Linear Programs (Version 1.0)," Technical Report OR80015, Department of Operations Research, Southern Methodist University, Dallas, 1980.

Armstrong, R., A. Charnes, and C. Haksever, "Successive Linear Programming for Ratio Goals," *European Journal of Operational Research*," 32 (1987), 426-434.

Armstrong, R., A. Charnes, and C. Haksever, "Implementation of Successive Linear Programming Algorithms for Non-Convex Goal Programming," *Computers and Operations Research*, 15 (1988), 37-49.

Charnes, A., W. M. Raike, J. D. Stutz, and A. S. Waters, " On Generation of Test Problems for Linear Programming Codes," *Communications of the ACM*, 17 (1974), 583-586.

Charnes, A., and W. W. Cooper, "Goal Programming and Multiple Objective Optimization, Part 1," *European Journal of Operational Research*, 1 (1977), 39-54.

Cheng, M. C., "Recent Development in the Simplex Algorithm," *Proceedings of the Mathematical Seminar: Singapore 1980*, K. C. Cheng and C. T. Chong (Eds.), Singapore Mathematical Society and Mathematics Department, N. U. S. Singapore, (1981) 6-19.

Cheng, M. C., "Generalized Theorems for Permanent Basic and Nonbasic Variables," *Mathematical Programming*, 31 (1985), 229-234.

Golden, B., L. Bodin, T. Doyle, and W. Stewart, Jr., "Approximate Travelling Salesman Algorithms," *Operations Research*, 28 (1980), 694-711.

Golden, B., and W. R. Stewart, "Empirical Analysis of Heuristics," in E. L. Lawler, J. K. Lenstra, A. H. G. Rinnooy Kan, and D. B. Shmoys (Eds.) *The Travelling Salesman Problem*, Wiley, Chichester, England, 1985.

Haksever, C. "An Implementation and Computational Evaluation of New Criteria for the Simplex Method," presented at the conference in honor of A. Charnes on his 70th birthday, October 14-15, 1987, IC^2 Institute, The University of Texas, Austin.

Haksever, C. and J. L. Ringuest, "Computational Experiments on the Efficient Implementation of Some MOLP Algorithms," Working Paper, 1989.

Haksever, C. and J. L. Ringuest, "Computational Efficiency and Interactive MOLP Algorithms: An Implementation of the SIMOLP Procedure," *Computers and Operations Research*, 17 (1989) pp. 39-50.

Keeney, R. L. and H. Raiffa, *Decisions with Multiple Objectives: Preferences and Value Tradeoffs*, Wiley, New York, 1976.

Lawler, E. L., J. K. Lenstra, A. H. G. Rinnooy Kan, and D. B. Shmoys (Eds.), *The Travelling Salesman Problem*, Wiley, Chichester, England, 1985.

Reeves, G. R., and L. S. Franz, "A Simplified Interactive Multiple Objective Linear Programming Procedure," *Computers and Operations Research*, 12 (1985), 589-601.

Rosenkrantz, D. J., R. E. Stearns, and P. M. Lewis, II, "An Analysis of Several Heuristics for the Travelling Salesman Problem," *SIAM Journal on Computing*, 6 (1974), 563-581.

Steuer, R. E., *Multiple Criteria Optimization: Theory, Computation, and Application*, Wiley, New York, 1986.

Zionts, S. and J. Wallenius, "An Interactive Multiple Objective Linear Programming Method for a Class of Underlying Nonlinear Utility Functions," *Management Science*, 29 (1983), 519-529.

USING MULTIOBJECTIVE LINEAR PROGRAMMING TO RECONCILE PREFERENCES OVER TIME

9.1   PREFERENCES OVER TIME

Decision makers are often faced with a choice among alternatives, each of which leads to a consequence $Z_j = (Z_{1j}, Z_{2j}, ..., Z_{kj})$, where Z_{hj} indicates the consequence in the hth year of choosing alternative j now. In general, the consequence in any given year may include nonmonetary factors as well as monetary ones. However, for purposes of this presentation, it will be assumed that all nonmonetary factors can be "costed out" so that Z_j can be considered a cash flow.

A first step in evaluating cash flows might be to assess a multiattribute value function over all possible streams. Bell (1974) and Meyer (1976) present alternative approaches for constructing these value functions. The use of value functions requires that the cash flows are known with certainty. Under uncertainty, multiattribute utility functions are appropriate. Because of the complexity involved in applying these methods, they are not in common use.

The usual procedure is to discount the cash flow using some standard interest rate or cost of capital. This is done using the net present value (NPV) formula:

$$NPV_j = \sum_{h=1}^{k} \frac{Z_{hj}}{(1+r)^{h-1}}$$

where in the certainty case r is the appropriate rate of return, and k is the number of periods. In the case of uncertainty, r is a risk-adjusted rate. The NPV function can, however, be viewed as a particular specification of the multiattribute value (utility) function. As such, there are a set of behavioral properties inherent in this form.

9.2   THE BEHAVIORAL PROPERTIES OF NPV

The behavioral properties implied by the net present value model are [Meyer (1976)]:

1. If all $Z_{h1} \geq Z_{h2}$ and at least one $Z_{h1} > Z_{h2}$, then Z_1 is preferred to Z_2 for all values of r, $0 < r < 1$. The converse, however, is not true; that is, when Z_1 is preferred to Z_2 it is not necessary that all $Z_{h1} \geq Z_{h2}$.

2. If Z_1 and Z_2 differ only in periods u and v [i.e. ($Z_{11} = Z_{12}, ..., Z_{u-1,1} = Z_{u-1,2}, Z_{u1} \neq Z_{u2}, Z_{u+1,1} = Z_{u+1,2}, ..., Z_{v-1,1} = Z_{v-1,2}, Z_{v1} \neq Z_{v2}, Z_{v+1,1} = Z_{v+1,2}, ..., Z_{k1} = Z_{k2})$], then the preference between Z_1 and Z_2 is dependent only on the returns in periods u and v and does not depend on the returns common to both Z_1 and Z_2.

3. For $h = 1, 2, ...$ one unit of return in period h is worth $1 + r$ units in period $h + 1$. Thus, there is a constant substitution rate between Z_{hj} and $Z_{h+1,j}$ and the contours of indifference curves for tradeoffs between Z_{hj} and $Z_{h+1,j}$ are parallel straight lines.

4. With r constant the indifference curves for tradeoffs between amounts Z_{hj} and $Z_{h+1,j}$, $h = 1, 2, ...$, have the same slope.

Two examples will serve to illustrate some of the difficulties inherent in these properties. Suppose that a decision maker could choose between alternatives which would yield the consequences Z_1 = (10, 16, 20, 25, 50, 100) or Z_2 = (10, 15, 20, 30, 50, 100), and that this particular decision maker prefers Z_1 to Z_2 due to the slightly higher returns in period 2 which would help with some temporary cash flow problems. Now suppose that the same decision maker in the same situation is presented with a choice between alternatives which yield the consequences Z_1 = (100, 16, 20, 25, 50, 100) or Z_2 =

(100, 15, 20, 30, 50, 100). With such high returns in period one, the decision maker has less concern about the short term cash flow problems of period two, and so prefers Z_2 to Z_1. This change in preference is inconsistent with property 2 of NPV.

Now consider the example of a decision maker with a discount rate r, who must choose between alternatives which yield the consequences, $Z_1 = (100, 100+100r, 100)$ or $Z_2 = (200, 0, 100)$, and suppose that the decision maker prefers Z_1 to Z_2 because of the smoother flow of returns which might be more comforting to stockholders. This choice is inconsistent with property 3 of the NPV model, which would require the decision maker to be indifferent between these two consequences.

These properties of the NPV formulation result directly from the functional form of the model. Net present value is an additive specification of the multiattribute value function. Each term in this additive specification is linear and the weight applied to each term is a function of a common discount rate, r. This is a simple and tractable, but restrictive representation.

Up to this point in the discussion, the cash flows have been treated as certain amounts. Future flows of returns are often uncertain, however, and should be treated as such. A simplistic approach for handling uncertainty is to use a risk-adjusted discount rate in the NPV formula. This risk-adjusted value could be the rate of return for previous decisions with comparable risk. In the absence of prior decisions of comparable risk, though, this risk-adjusted rate may be difficult to establish. The risk-adjusted model is a particular specification of a multiattribute utility function with behavioral assumptions analogous to those presented for the certainty case.

9.3   A MORE GENERAL NPV MODEL

The simplest way that the NPV model may be generalized is by using a variable discount rate. This requires that the NPV formula be modified by replacing the constant discount rate (or risk-adjusted discount rate), r, with a variable discount rate (or variable risk-adjusted discount rate), r_h. A distinct estimate of r_h is obtained for each h. With a variable discount rate, property 1 above still holds. Properties 2 and 3 are modified as follows:

2m.   If Z_1 and Z_2 differ only in periods h and $h+1$ [i.e. $(Z_{11}=Z_{12}, ..., Z_{h-1,1}=Z_{h-1,2}, Z_{h1} \neq Z_{h2}, Z_{h+1,1} \neq Z_{h+1,2}, Z_{h+2,1}=Z_{h+2,2}, ..., Z_{k1}=Z_{k2})$], then the preference between Z_1 and Z_2 is dependent only on the returns in periods h and $h+1$ and does not depend on the returns common to both Z_1 and Z_2.

3m.   For $h=1, 2, ...$, one unit of return in period h is worth $1+r_h$ units in period $h+1$; thus, there is a constant substitution rate between Z_h and Z_{h+1} and the contours of indifference curves for tradeoffs between amounts Z_{hj} and $Z_{h+1,j}$ are parallel straight lines.

Property 4 no longer holds. Thus, the variable discount rate yields only a slight generalization of the properties of the NPV model.

The two examples above would still be applicable although in slightly modified form. In the first example, the only differences between the two consequences would have two occur in consecutive time periods. In the second example, r would simply be replaced by r_h. These are still meaningful restrictions on decision maker behavior.

One way to further generalize the approach to reconciling preferences over time is to construct multiattribute value (or utility) functions which more properly represent the decision maker's time preferences and risk attitudes. Such functions can be constructed using the methods of Bell (1974) or Meyer (1976). Their approaches typically yield nonlinear functional forms for the multiattribute value (utility) functions.

The methods described to this point can be used when either a finite or infinite set of decision alternatives exist. When there are infinitely many decision alternatives, the choice among them becomes an optimization problem where the objective function is the multiattribute value (utility) function. This objective is usually represented by the NPV function because it is easier to optimize than a nonlinear representation. In some cases, however, a multiobjective formulation can be used. The multiobjective formulation is straightforward and requires no behavioral assumptions.

9.4   USING MULTIOBJECTIVE LINEAR PROGRAMMING AS AN ALTERNATIVE TO NPV

If continuous variables, x, can be used to model the decision alternatives (e.g., dollars invested in stocks, R&D or capital expenditures) which yield the cash flows, $Z(x)$, the problem can be represented by a multiobjective math program. Specifically, the problem is:

Maximize $\{Z(x) \mid x \, R\}$

where R is the domain of x. Further, if $R = \{x \mid g_i(x) = 0; i = 1, ..., m\}$ and $g(x)$ is a set of m linear constraints, the problem is a multiple objective linear program. $g(x)$ includes the cash flow restrictions and any other constraints on x which are typical of multi-period investment problems.

Any multiple objective simplex algorithm [e.g. Evans and Steuer (1973), Philip (1972), or Zeleny (1974)] can be used to solve this problem. The result will be a set of nondominated cash flow vectors, $Z(x)$, and their corresponding efficient alternative vector(s), x.

The application of dominance (or first order stochastic dominance) criteria is a common first step in decision analysis. In this case using multiobjective linear programming reduces the infinite set of cash flows to a set of nondominated solutions. There are an infinite number of nondominated solutions if all of the nondominated faces and edges are considered but as discussed in Chapter 5 these are just weighted linear combinations of the nondominated extreme points. The techniques described in Chapter 5 (i.e. filtering, clustering, sectioning, matching and grouping, and stochastic screening) can be used to reduce the set of nondominated solutions to a more manageable number.

Once a small set of candidate cash flows (nondominated solutions) has been obtained it may be possible to use what Meyer (1976) calls generalized net present value. This method requires the decision maker to collapse each cash flow vector to a single value. The process is a sequential one in which the decision maker replaces the vector $Z_j = (Z_{1j}, ..., Z_{kj})$, with another vector $Z'_j = (Z_{1j}, ..., Z'_{k-1,j}, 0)$ such that he/she is indifferent between the two. Next, the decision maker constructs a vector $Z''_j = (Z_{1j}, ..., Z'_{k-2,j}, 0, 0)$ so that he/she is indifferent between Z'_j and Z''_j. This process is continued until an equivalent cash flow vector $(Z'_{1j}, 0, ..., 0)$ is found. This single value, Z'_{1j}, may then be compared to those obtained from other nondominated solutions. Thus, it will not be necessary to specify or optimize a multiattribute value (utility) function.

One desirable property of these nondominated solutions is that the maximum NPV solution is always included among them. This can be shown as follows:

Theorem. The solution Z^* which yields the maximum net present value is nondominated for all r_h, $0 < r_h < 1$.

Proof. Suppose that x^* is dominated, then there is another solution x such that:

and
$$Z_h \geq Z^*_h \quad \text{for all h}$$
$$Z_h > Z^*_h \quad \text{for some h.}$$

Then it should also be true that

$$\sum_{h=1}^{k} \frac{Z_h}{(1+r_h)^{h-1}} > \sum_{h=1}^{k} \frac{Z^*_h}{(1+r_h)^{h-1}}.$$

But, this cannot be true, therefore, Z^* is nondominated.

9.5 THE ADVANTAGES OF USING MULTIOBJECTIVE LINEAR PROGRAMMING FOR RECONCILING PREFERENCES OVER TIME

Through the sequence of steps: 1) solution of the multiobjective linear program to produce a list of nondominated solutions, 2) if necessary, reducing the number of nondominated solutions using methods such as those described in Chapter 5, and 3) evaluating the remaining alternatives perhaps using generalized net present value, the decision maker is led to an unambiguous choice. The multiobjective problem in step 1 is well defined mathematically and is completely objective. In steps 2 and 3, the decision maker can use any means or decision aids to choose an alternative from the set of nondominated solutions. Both quantitative and qualitative information can be considered in making this choice and no discount rates or value (utility) functions need be specified.

9.6 REFERENCES

Bell, D. E., "Evaluating Time Streams of Income", *OMEGA*, 2 (1974), 691-699.

Evans, J. P. and R. E. Steuer, "A Revised Simplex Method for Linear Multiobjective Programs', *Mathematical Programming*, 5 (1973), 54-72.

Meyer, R. F., "Preferences Over Time". In R. L. Keeney and H. Raiffa, *Decisions with Multiple Objectives: Preferences and Value Tradeoffs*, Wiley, New York, 1976.

Philip, J., "Algorithms for the Vector Maximization Problem", *Mathematical Programming*, 2 (1972), 207-229.

Zeleny, M., *Linear Multiobjective Programming*, Springer-Verlag, New York, 1974.

10

DATA PRESENTATION AND MULTIOBJECTIVE OPTIMIZATION

Earlier chapters presented a number of methods for conducting multiobjective optimization. These methods frequently require the active participation of the decision maker. This means that the information provided to and elicited from the decision maker must be in a form that the decision maker can understand and use. Studies have shown that data representation can affect decision maker responses. Thus, the form in which data is presented to the decision maker is important.

10.1 DATA REPRESENTATION AND THE AXIOMS OF UTILITY THEORY

Some research has been conducted on the effects of data representation on a decision maker's compliance with the axioms of utility theory. Moskowitz (1974) studied alternative problem representations in the context of the Sure-Thing Principle. Keller (1982, 1983, and 1985) investigated the effects of three forms of problem representation on decision maker's responses to problems testing the Sure-Thing and Substitution Principles.

10.1.1 Data Representation and the Sure-Thing Principle

One axiom of utility theory is the principle that choices between alternatives should not depend on an event for which each alternative has the same "sure-thing" outcome. Moskowitz (1974) presented subjects with decision trees, matrices and written statements on three Allais-Paradox problems. As an example of an Allais problem, suppose a decision maker is faced with four options:

Option 1 =   100% chance of $1 million

Option 2 ≡   ⎡ 10% chance of $5 million
                    ⎢ 89% chance of $1 million
                    ⎣ 1% chance of $0

Option 3 ≡   ⎡ 10% chance of $5 million
                    ⎣ 90% chance of $0

Option 4 ≡   ⎡ 11% chance of $1 million
                    ⎣ 89% chance of $0

The decision maker has two decisions: 1) the decision maker must choose between options 1 and 2, and 2) between options 3 and 4. Choosing option 2 over option 1 would result in both reducing the probability of a $1 million payoff by 0.11 and increasing the probability of $5 million and $0 payoffs by 0.1 and 0.01, respectively. Choosing option 3 over option 4 would yield the identical changes (i.e. both reducing the probability of a $1 million payoff by 0.11 and increasing the probability of $5 million and $0 payoffs by 0.1 and 0.01, respectively). The Allais-Paradox is that most decision makers will choose option 1 and option 3 which is inconsistent. Moskowitz found that presenting these problems as decision trees led to more violations than when the problems were presented as matrices or written statements.

Keller (1982, 1983 and 1985) also examined three forms of problem representation on the Sure-Thing Principle. The first was a written problem statement of the form:

Option 1 =   100% chance of $3000

Option 2 ≡ $\left\{\begin{array}{l} 75\% \text{ chance of } \$3000 \\ 5\% \text{ chance of } \$0 \\ 20\% \text{ chance of } \$4000. \end{array}\right.$

The second form was a picture of marbles in tubes. In this form problems were represented as a choice between two tubes, each containing 100 labeled marbles as in Figure 10.1A. Here the "sure-thing" payoff of $3000 is partitioned off by the heavy line. The third representation of the problem was a proportional decision matrix as shown in Figure 10.1B. In this form the width of a column is proportional to the probability of the corresponding outcome. Keller found that the proportional matrix representation resulted in fewer violations of the Sure-Thing Principle than either of the other representations.

10.1.2 Data Representation and the Substitution Principle

Keller (1982, 1983 and 1985) also studied the effect of problem representation on violations of the Substitution Principle. The Substitution Principle can be described by the following example:

Option 1 ≡ $\left\{\begin{array}{l} 80\% \text{ chance of } \$4000 \\ 20\% \text{ chance of } \$0 \end{array}\right.$

Option 2 =   100% chance of $3000

Option 3 ≡ $\left\{\begin{array}{l} 20\% \text{ chance of } \$4000 \\ 80\% \text{ chance of } \$0 \end{array}\right.$

Option 4 ≡ $\left\{\begin{array}{l} 25\% \text{ chance of } \$4000 \\ 75\% \text{ chance of } \$0. \end{array}\right.$

Figure 10.1

A. Tubes

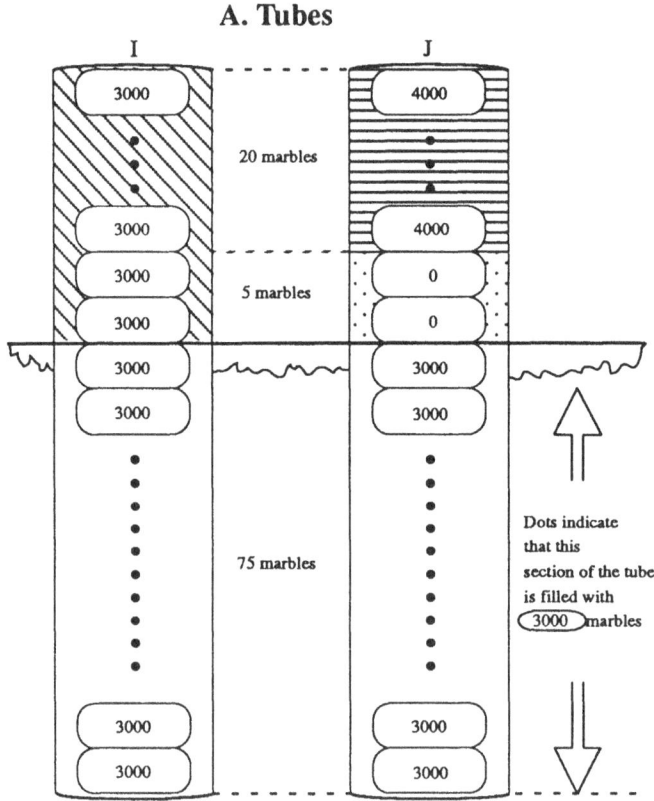

B. Proportional Matrices

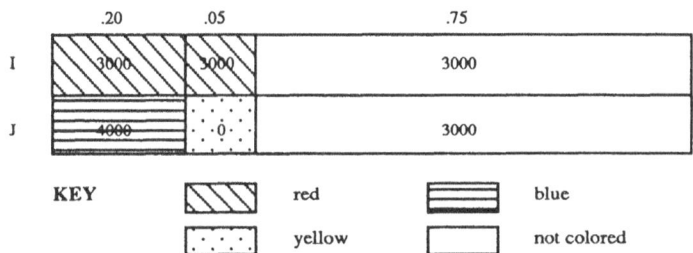

Reprinted by permission of L. Robin Keller, "The Effects of Problem Representation on the Sure-Thing and Substitution Principles," *Management Science*, Volume 31, Number 6, June 1985. Copyright 1985, The Institute of Management Sciences.

Notice that options 3 and 4 are equivalent to the following:

Option 3 $\equiv$ $\begin{cases} 25\% \text{ chance of Option 1} \\ 75\% \text{ chance of \$0} \end{cases}$

Option 4 $\equiv$ $\begin{cases} 25\% \text{ chance of Option 2} \\ 75\% \text{ chance of \$0.} \end{cases}$

Thus, if option 1 is preferred to option 2 the Substitution Principle requires that option 3 be preferred to option 4. Keller tested the effect of the three problem representations described in the previous section on violations of this principle. The results obtained again showed fewer violations using proportional matrices.

While the information elicited from the decision makers in these studies is not precisely the same information required in a multiobjective optimization problem, these studies do show that the way a problem is presented can affect a decision maker's ability to choose rational alternatives. Winkels and Meika (1984), Korhonen and Laakso (1986) and Korhonen (1987, 1988) have incorporated graphical data presentations in multiobjective settings. However, to date no experimental comparison of alternative data presentations has been conducted in this context.

In each of the studies and examples described above the decision maker was presented with the same information in several different pictorial forms. Perhaps more important than the depiction of the data is the frame used for the decision.

10.2   THE FRAMING OF DECISIONS

Tversky and Kahneman (1981) use the term "decision frame" to describe "the decision maker's conception of the acts, outcomes and contingencies that make up a particular choice." The particular frame that a decision maker employs is determined in part by the decision maker's personal characteristics and partly by the formulation of the problem. It is very often possible for the same decision maker to frame a given problem in more than one way. Tversky and Kahneman compare alternative frames to alternative perspectives on a visual scene. Ideally visual perception would require the relative heights of two neighboring objects to remain constant with changes of vantage point. Similarly, it would be ideal for preferences between options to remain constant with changes of decision frame. However, because of imperfections of human perception and decision making, neither visual perception nor preferences remain constant. Tversky and Kahneman have observed systematic reversals of preferences in a variety of problems and in the choices of different groups of respondents.

10.2.1 Reference Effects

In multiobjective optimization the decision maker is often faced with a variety of outcomes. Outcomes are commonly perceived as positive or negative in relation to a reference outcome that is considered neutral. Varying the reference point could therefore determine whether a particular outcome is perceived as positive or negative. Further, Tversky and Kahneman have found that in general value functions are concave for gains and convex for losses. That is, most people are risk averse when faced with a choice involving gains, but, risk prone when faced with a choice involving losses. In addition, they observe that the value function is steeper for losses than for gains. If a value function is steeper for losses than gains, a difference between outcomes will seem greater when the outcomes are framed as losses rather than gains.

As an example of these effects, credit-card industry representatives have proposed framing a proposal to pass to the consumer some of the costs associated with the processing of credit-card purchases as a cash discount rather than a credit-card surcharge. The two frames result in different reference points (i.e. a higher price with a discount for cash versus a lower price with a surcharge for credit) suggesting a reference to the higher or the lower of the two prices. Again, because in Tversky and Kahneman's words, "losses loom larger than gains," consumers are less willing to accept a surcharge than to forgo a discount.

The goal programming formulation of the multiobjective optimization problem is particularly susceptible to reference effects as deviations from a goal will be positive or negative based solely on the aspiration level set for the goal. By setting the aspiration level high, which is often done to insure that the solutions obtained are nondominated (assuming the objective is to be maximized), the goal will be underachieved. This underachievement is likely to be weighted more heavily by the decision maker than would an overachievement of equal magnitude. Thus, any trade-offs among objectives will be influenced by the aspiration levels specified for each goal and whether the goals are underachieved or overachieved.

10.2.2 Sunk-Cost Effects

In many situations a decision results in an outcome which becomes part of a series of changes in a single attribute that are brought about not only by the current decision but by other decisions as well. For example, a sequence of monetary gains or losses that are the result of a series of related or unrelated decisions. In other cases a decision results in a set of concurrent changes in several attributes. Tversky and Kahneman describe the framing and evaluation of these compound outcomes using the notion of a psychological account. This account is a frame which specifies the set of outcomes that make up the compound outcome, the manner in which these outcomes are combined, and a reference outcome that is considered neutral. They further propose that most people evaluate compound outcomes in terms of a minimal account, which includes only the direct consequences of the outcome. The rationale for the notion of minimal accounts is that this frame simplifies evaluation and reduces cognitive strain, reflects the intuition that consequences should be linked to decisions, and is more sensitive to changes both desirable and undesirable than to steady state.

In some situations, however, the outcomes of a decision affect an account that was previously established by some related decision. In these cases, the current decision may be evaluated in terms of a more detailed account. When a decision is referred to an existing account in which the current balance is negative a sunk-cost effect often arises. This effect is illustrated by the devaluation of money which often results in extra spending or by a reduction in the significance of small discounts in the context of a large expenditure. Because of the nonlinearities of the evaluation process, the minimal account and a more detailed account often lead to different choices.

The objectives in mathematical programming problems frequently measure only marginal effects. For example, the maximization of contribution to profit is a common objective. In making tradeoffs among objectives the sunk-cost effect may arise.

The goal programming formulation will again be susceptible to this effect since in goal programming, tradeoffs are made between deviations from specified aspiration levels for each objective.

10.3   RECONCILING THE DECISION FRAME

The susceptibility of decision making to perspective effects is of special concern because of the absence of objective standards for comparing frames.   Individuals who face decisions are normally unaware of alternative frames and of their potential effects.   They would like their preferences to be independent of frames, but they are often uncertain how to resolve inconsistencies.   In some cases the advantage of one frame over another is obvious once the competing frames are compared, but in many other cases it is not clear which frames are inferior.

The practice of acting on the most readily available frame is sometimes justified due to the mental effort required to explore alternative frames. However, the potential pitfalls from not analyzing alternative frames may offset the cost of thinking. A predictive orientation in decision making encourages a focus on future experience. In particular, predictive considerations may be used to select the decision frame that best represents the future experience of outcomes.

The framing of decisions and outcomes can also reflect the acceptance or rejection of responsibility for particular consequences. Further, by manipulating the decision frame, the perception of an outcome can be controlled. Thus, the adoption of a decision frame is ethically significant.

10.4   PERCEPTION OF THE IDEAL

As discussed in Chapter 4, Zeleny (1982) argues that a rational decision maker will chose the alternative which is deemed closest to the perceived ideal. In compromise programming it is often possible to specify a single (usually unobtainable) point-valued estimate of the ideal solution. However, this is not always the case since the feasible solution space may be unbounded in the direction of one or more objectives.

When a point-valued estimate of the ideal solution cannot be obtained one must deal with what Zeleny refers to as a fuzzy-valued estimate. The ideal solution is not then viewed by the decision maker as a point but as a "hazy cloud," [Zeleny (1982)]. This haziness will have a substantial impact on the decision maker's ability to discriminate among alternatives and will ultimately impact the final decision.

Fuzziness in the definition of the ideal solution may also be confused with displacement of the ideal solution as described in Section 4.5. Further, if a decision maker can be persuaded that the ideal has been displaced when it is only hazily specified, the decision maker may be persuaded to make choices which would not be made if the ideal were precisely specified. Thus, manipulation of the perceived ideal can impact the decision maker's final choice in much the same way as manipulation of the decision frame.

10.5   REFERENCES

Keller, L. R., "The Effects of Problem Representation on Conformity with Utility Properties - An Empirical Investigation," unpublished doctoral dissertation, University of California, Los Angeles, Graduate School of Management, 1982.

Keller, L. R., "The Effects of Problem Representation on the Sure-Thing and Substitution Principles," Working Paper, University of California, Irvine, Graduate School of Management, 1983.

Keller, L. R. "The Effects of Problem Representation on the Sure-Thing and Substitution Principles," *Management Science*, 31 (1985), 738-751.

Korhonen, P., and J. Laakso, "A Visual Interactive Method for Solving the Multiple Criteria Problem," *European Journal of Operational Research*, 24 (1986), 277-287.

Korhonen, P., "VIG - A Visual Interactive Support System for Multiple Criteria Decision Making," *Belgian Journal of Operations Research, Statistics and Computer Science*, 27 (1987), 3-15.

Korhonen, P., "A Visual Reference Direction Approach to Solving Discrete Multiple Criteria Problems," *European Journal of Operational Research*, 34 (1988), 152-159.

Moskowitz, H., "Effects of Problem Representation and Feedback on Rational Behavior in Allais and Morlat-Type Problems," *Decision Sciences*, 5 (1974), 225-242.

Tversky, A. and D. Kahneman, "The Framing of Decisions and the Psychology of Choice," *Science*, 211 (January 30, 1981), 453-458.

Winkels, H. M. and M. Meika, "An Integration of Efficiency Projections Into the Geoffrion-approach for Multi-objective Linear Programming," *European Journal of Operational Research*, 16 (1984), 113-127.

Zeleny, M., *Multiple Criteria Decision Making*, Mcgraw-Hill, New York, 1982.

INDEX

The manufacturer's authorised representative in the EU is Springer
Nature Customer Service Centre GmbH, Europaplatz 3, 69115 Heidelberg,
Germany. If you have any concerns regarding our products, please
contact ProductSafety@springernature.com

Printed and bound by CPI Group (UK) Ltd, Croydon, CR0 4YY

24/04/2026

02096348-0002